Digital Logic Design

MANSAF ALAM
Assistant Professor
Department of Computer Science
Faculty of Natural Sciences
Jamia Millia Islamia, New Delhi

Editor-in-Chief
Journal of Applied Information Science

BASHIR ALAM
Assistant Professor
Department of Computer Engineering
Faculty of Engineering and Technology
Jamia Millia Islamia, New Delhi

PHI Learning Private Limited

Delhi-110092
2016

₹ 295.00

DIGITAL LOGIC DESIGN
Mansaf Alam and Bashir Alam

© 2016 by PHI Learning Private Limited, Delhi. All rights reserved. No part of this book may be reproduced in any form, by mimeograph or any other means, without permission in writing from the publisher.

ISBN-978-81-203-5108-0

The export rights of this book are vested solely with the publisher.

Published by Asoke K. Ghosh, PHI Learning Private Limited, Rimjhim House, 111, Patparganj Industrial Estate, Delhi-110092 and Printed by Raj Press, New Delhi-110012.

To
Our parents, family, friends and students

Contents

Preface xiii

Acknowledgement xv

1. Introduction to Digital System 1–10

    1.1 Introduction *1*
    1.2 Digital *1*
    1.3 Digital System *2*
        1.3.1 Digital Electronics *2*
        1.3.2 Digital Signal Processing (DSP) *2*
    1.4 Design Issues in Digital Circuit *3*
    1.5 Digital Computer *3*
        1.5.1 Digital Computer Development *3*
        1.5.2 Digital Computer Functional Element *3*
    1.6 Specification and Implementation of Digital Design *4*
        1.6.1 Specification *5*
        1.6.2 Implementation *5*
    1.7 Structured and Trial Methods in Design *6*
    1.8 CAD Tools *6*
        1.8.1 CAD Tools Available for Designing Digital Logic Circuit *7*
    Exercises *8*
        Short Answer Questions *8*
        Long Answer Questions *8*
        Multiple Choice Questions *9*
    Bibliography *10*

2. Number Systems 11–40

    2.1 Introduction *11*
    2.2 Definition *11*
    2.3 Binary Number *11*
    2.4 Octal Number *12*
    2.5 Hexadecimal Number *12*
    2.6 Hexal Number *13*
    2.7 Nano Number *13*

2.8 Decimal to Binary Conversion *14*
    2.8.1 The Conversional of Floating Point Decimal Number to Binary Number *14*
    2.8.2 Direct Method *16*
2.9 Binary to Decimal Conversion *16*
    2.9.1 Binary Fraction Number to Decimal Conversion *17*
2.10 Decimal to Octal Conversion *18*
    2.10.1 Decimal Fraction to Octal Conversion *18*
2.11 Octal to Decimal Conversion *20*
    2.11.1 Octal Fraction to Decimal Conversion *21*
2.12 Octal to Binary Conversion *22*
    2.12.1 Octal Fraction to Binary Conversion *22*
2.13 Binary to Octal Conversion *24*
    2.13.1 Binary Fraction to Octal Conversion *24*
2.14 Decimal to Hexadecimal Conversion *25*
    2.14.1 Decimal Fraction to Hexadecimal Conversion *26*
2.15 Hexadecimal to Decimal Conversion *27*
    2.15.1 Hexadecimal Fraction to Decimal Conversion *27*
2.16 Hexadecimal to Binary Conversion *28*
    2.16.1 Hexadecimal Fraction to Binary Conversion *29*
2.17 Binary to Hexadecimal Conversion *31*
    2.17.1 Binary Fraction to Hexadecimal Conversion *31*
2.18 Decimal to Hexal Number Conversion *32*
    2.18.1 Decimal Fraction to Hexal Conversion *32*
2.19 Hexal Number to Decimal Number Conversion *35*
    2.19.1 Hexal Fraction to Decimal Conversion *35*
2.20 Decimal Number to Nonal Number Conversion *36*
    2.20.1 Decimal Fraction to Nonal Conversion *36*
2.21 Nonal Number to Decimal Number Conversion *37*
    2.21.1 Nonal Fraction to Decimal Conversion *37*
Exercises *38*
    Short Answer Questions *38*
    Long Answer Questions *38*
    Multiple Choice Questions *39*
Bibliography *40*

3. Data and Information Representation 41–64

3.1 Introduction *41*
3.2 Data and Information *41*
3.3 Number Normalization *41*
3.4 Floating Point Representation *43*
    3.4.1 Floating Point Representation of 32 Bits Binary Number *43*
    3.4.2 Floating Point Representation of 16 Bits Binary Number *44*
    3.4.3 Range of 16 Bits Binary Number *45*
    3.4.4 Range of 32 Bits Binary Number *46*

3.5 Representation of Negative Number   47
    3.5.1 Representation of Negative Number in Binary System   47
    3.5.2 Representation of Negative Number in Octal System   48
    3.5.3 Representation of Negative Number in Decimal System   49
    3.5.4 Representation of Negative Number in Hexadecimal System   51
3.6 Codes and Its Conversion   52
    3.6.1 BCD Code   52
    3.6.2 Gray Code   52
    3.6.3 Excess-3 Code   53
    3.6.4 8421 Code   54
    3.6.5 2421 Code   55
    3.6.6 $8\overline{4}21$ Code   55
    3.6.7 ASCII Code   56
    3.6.8 Self-complement Code   56
3.7 Error Detection and Correction Code   56
    3.7.1 Odd Parity Method   56
    3.7.2 Even Parity Method   57
    3.7.3 Hamming Code Method   57
    3.7.4 Parity Bit Position   58
    3.7.5 Error Position in Message   59
Exercises   61
    Short Answer Questions   61
    Long Answer Questions   61
    Multiple Choice Questions   62
Bibliography   64

4. Computer Arithmetic   65–90

4.1 Computer Arithmetic   65
4.2 Binary Arithmetic   65
    4.2.1 Binary Addition   65
    4.2.2 Binary Subtraction   66
    4.2.3 Binary Multiplication   69
    4.2.4 Binary Division   69
4.3 Octal Arithmetic   70
    4.3.1 Octal Addition   70
    4.3.2 Octal Subtraction   70
4.4 Decimal Arithmetic   73
4.5 Hexadecimal Arithmetic   75
    4.5.1 Addition   75
    4.5.2 Hexadecimal Subtraction   76
4.6 Hexal Arithmetic   78
    4.6.1 Hexal Addition   78
    4.6.2 Hexal Subtraction   79

            4.6.3   Hexal Multiplication   79
            4.6.4   Hexal Division   80
      4.7   Nonal Arithmetic   83
            4.7.1   Nonal Addition   83
            4.7.2   Nonal Subtraction   83
            4.7.3   Nonal Multiplication   86
            4.7.4   Nonal Division   87
      Exercises   *87*
            Short Answer Questions   *87*
            Long Answer Questions   *87*
            Multiple Choice Questions   *89*
      Bibliography   *90*

5. Fundamentals of Boolean Logic and Gates 91–118

   5.1   Introduction   *91*
         5.1.1   Power Set   *91*
   5.2   Examples   *92*
         5.2.1   Boolean Algebra   *92*
   5.3   Postulate   *93*
   5.4   Theorem   *93*
   5.5   Boolean Function   *94*
   5.6   Logic Gates   *96*
         5.6.1   AND Logic Gate   *96*
         5.6.2   OR Logic Gate   *96*
         5.6.3   NOT Logic Gate   *97*
         5.6.4   NAND Gate   *98*
         5.6.5   NOR Gate   *99*
         5.6.6   Exclusive-OR Gate   *100*
         5.6.7   Exclusive-NOR Gate   *100*
   5.7   Circuit of Boolean Function   *101*
   5.8   NAND Gate Implementation of Boolean Function   *102*
         5.8.1   Procedure for NAND Gate Implementation   *103*
   5.9   NOR Gate Implementation of Boolean Function   *106*
         5.9.1   Procedure for NOR Gate Implementation   *107*
   Exercises   *112*
         Short Answer Questions   *112*
         Long Answer Questions   *112*
         Multiple Choice Questions   *116*
   Bibliography   *118*

6. Simplification of Boolean Function 119–133

   6.1   Introduction   *119*
   6.2   Simplification by Boolean Algebra   *119*

6.3 Canonical Forms of Boolean Algebra *120*
6.4 Simplification by Karnaugh-map (K-map) *121*
6.5 Tabular Method for Simplification *127*
Exercises 129
  Short Answer Questions 129
  Long Answer Questions 129
  Multiple Choice Questions 131
Bibliography 133

7. Combinational Circuit Design   134–179

7.1 Introduction *134*
7.2 Half Adder *136*
7.3 Full Adder *137*
7.4 N-bit Parallel Adder *138*
  7.4.1 Four-bit Parallel Adder *139*
7.5 Half Subtractor *139*
7.6 Full Subtractor *140*
7.7 Four-bit Parallel Subtractor *142*
  7.7.1 N-bit Parallel Subtractor *143*
7.8 Multiplexers *143*
  7.8.1 4 × 1 Multiplexer *144*
  7.8.2 8 × 1 Multiplexer *146*
7.9 Implementation of Multiplexer using K-map *150*
7.10 Demultiplexer *155*
7.11 Decoder *155*
7.12 Encoder *165*
7.13 Magnitude Comparator *166*
7.14 Read Only Memory (ROM) *169*
  7.14.1 Combinational Circuit Implementation with ROM *170*
7.15 Programmable Logic Array (PLA) *172*
7.16 Programmable Array Logic (PAL) *175*
Exercises 175
  Short Answer Questions 175
  Long Answer Questions 176
  Multiple Choice Questions 177
Bibliography 179

8. Sequential Circuit Design   180–208

8.1 Introduction *180*
8.2 Flip Flops *180*
8.3 Types of Flip Flops *181*
8.4 Sequential Circuit Design Procedure *187*
  8.4.1 Design Procedure in Case of State Equation *187*

        8.4.2 Design Procedure in Case of State Table   *190*
        8.4.3 Design Procedure in Case of State Diagram   *193*
    8.5 Random Access Memory (RAM)   *200*
    8.6 Algorithmic State Machine (ASM)   *200*
Exercises   202
    Short Answer Questions   202
    Long Answer Questions   203
    Multiple Choice Questions   205
Bibliography   208

9. Counter Design   209–232

    9.1 Counter   *209*
        9.1.1 Applications of Counter   *209*
    9.2 Binary Counter   *210*
    9.3 Synchronous Up/Down Counter   *210*
    9.4 Unused States   *213*
    9.5 Synchronous BCD Counter   *215*
    9.6 Decade Counter   *218*
    9.7 Ring Counter   *220*
    9.8 Johnson Counter   *223*
    9.9 Ripple Counter   *226*
Exercises   228
    Short Answer Questions   228
    Long Answer Questions   229
    Multiple Choice Questions   230
Bibliography   232

10. Register Design   233–243

    10.1 Introduction   *233*
    10.2 Types of Register   *234*
    10.3 Adding a Parallel Load Operation   *234*
    10.4 Register with Parallel Load   *235*
    10.5 Shift Register   *235*
        10.5.1 Serial In Serial Out (SISO) Shift Register   *236*
        10.5.2 Serial In Parallel Out (SIPO) Shift Register   *237*
        10.5.3 Parallel In Serial Out (PISO) Shift Register   *238*
        10.5.4 Parallel In Parallel Out (PIPO) Shift Register   *239*
        10.5.5 Bidirectional Shift Register (BSR)   *240*
Exercises   240
    Short Answer Questions   240
    Long Answer Questions   241
    Multiple Choice Questions   241
Bibliography   243

11. Threshold Circuit and Digital Computer Design   244–268

    11.1 Threshold Logic Circuit   *244*
        11.1.1 Threshold Gate (T-gate)   *244*
        11.1.2 Implementation of Conventional Gates with T-gate   *247*
    11.2 Arithmetic Circuit Design   *251*
        11.2.1 Arithmetic Circuit   *252*
        11.2.2 Logical Circuits   *257*
        11.2.3 Arithmetic Logic Unit (ALU)   *257*
    11.3 Central Processing Unit (CPU)   *260*
    Exercises   264
       Short Answer Questions   264
       Long Answer Questions   265
       Multiple Choice Questions   266
    Bibliography   268

Appendix: Miscellaneous Objective Type Questions   *269–273*

Index   *275–280*

Preface

The digital logic design is an interesting course to teach and study. In this book we have focused on the concepts in digital logic design. We feel that the basic concepts of digital logic design should be clear to the students while reading any book. The contents of this book have been organized in a systematic manner so as to inculcate sound knowledge and concepts amongst its readers.

Chapters 1–4 are introductory chapters and are meant for readers who are new to digital logic. These chapters cover basic concepts in combinational and sequential circuit design such as Digital electronics, Digital signal processing, Number systems, Data and information representation and, Computer arithmetic. Chapters 5–8 cover topics like Fundamentals of Boolean logic and gates, Simplification of Boolean function, Combinational circuit design and Sequential circuit design. Chapters 9–11 discuss advanced topics in digital logic design such as various types of Counter design, Register design, ALU design, Threshold circuit and digital computer design.

Latest concepts are introduced in this book that motivates the readers to learn these newer concepts through this book. All the concepts are explained with suitable examples which will be beneficial to both the teachers as well as students. In this book we have tried to explain the concepts of digital logic in an easy to understand manner to reinforce the understanding of the subject matter. Introduction to latest tools in digital design is given in this book which will help its readers to design circuits under different conditions. The extensive use of graphs and diagrams can develop better learning skills amongst teachers and students who are graphic learners. We have also considered such type of learners while writing this book. The book can be used by teachers and students for digital logic courses taught at diploma, undergraduate and postgraduate level. It covers the syllabus of digital logic/digital logic design courses offered in almost all universities.

In order to help students prepare them for competitive examinations such as GATE, NET, Engineering services, JTO, and university exams for digital logic course, short, long and multiple choice type questions are given at the end of each chapter. Apart from this a Question Bank containing numerous multiple choice questions is also given at end of this book covering all the chapters.

We welcome all the comments and suggestions for the improvement in this book.

MANSAF ALAM
BASHIR ALAM

Acknowledgement

We would like to extend our gratitude towards Jamia Millia Islamia, New Delhi, for providing us conducive and congenial atmosphere which proved to be a driving force that led us to write this book.

We sincerely thank our Vice Chancellor, Prof. Talat Ahmad, Former Dean, Prof. Sharfuddin Ahmad and Prof. M.N. Doja, Founder Head, Department of Computer Engineering, for their constant, encouragement and support.

We acknowledge the encouragement provided to us by our colleagues at Departments of Computer Science, Computer Engineering and Mathematics, Jamia Millia Islamia, during the course of writing this book. We especially thank Dr. Israr Ahmad, Mr. Abdul Aziz, Mr. Ziaullah Siddiqui and Mr. H.M. Kidwai for their moral support throughout the writing of this book.

We also thank our students and research scholars for their valuable feedback and suggestions on the book. We would also like to acknowledge the continuous help, suggestions, criticism and support extended to us by Ms. Kashish Ara Shakil and Dr. Khalid Raza while writing this manuscript.

We thank the entire team of PHI Learning for efficiently and smoothly handling all the phases for publication of this book.

Lastly, we would like to thank our friends including Dr. Arshad Khan, Dr. M.Y. Abbasi, Dr. Shahzad Hasan and Dr. S.K. Naqvi and our family members including Mr. Md. Ayoob and Mr. Khalid Hasan for constantly encouraging and motivating us.

MANSAF ALAM
BASHIR ALAM

CHAPTER 1

Introduction to Digital System

1.1 INTRODUCTION

The modern digital computers are based on digital logic circuit. To know how computer system works, you have to understand digital logic and Boolean algebra. This chapter provides only a basic introduction to the components used in computer system design.

1.2 DIGITAL

The term digital means data are represented in high voltage and low voltage. The high voltage is represented by 1 and low voltage is represented by 0. In general, digit signals are represented by only two possible values 1 and 0. In other words you can say true in case of high voltage and false in case of low voltage. Many complex signals can be constructed by using 1s and 0s, end-to-end, like a necklace. If you use four binary digits end-to-end, you have sixteen possible combinations as 0000, 0001, 0010, 0011, 0100, 0101, 0110, 0111, 1000, 1001, 1010, 1011, 1100, 1101, 1110 and 1111. There is no limit of binary digit, you can use many binary digits to form signal as per your requirement. A typical digital signal is shown in Figure 1.1, in which firstly represented as a series of voltage levels that change as time goes on, and then as a series of 1s and 0s.

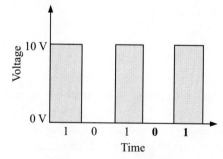

Figure 1.1 A digital signal.

1.3 DIGITAL SYSTEM

Digital system processes the digital data and produces the digital information. The digital system is operated on discrete data, which is obtained from counting. The digital system can process the digital data only. The analogue data or continuous data, which is obtained through measurement, are not processed by digital system. The examples of digital system are two broad categories of digital system namely digital electronics and digital signal processing.

1.3.1 Digital Electronics

Digital electronic is also known as digital circuits which designed with logic gate like AND, OR, NOT, NAND and NOR. It represents signals by digital bands of analogue levels, rather than an analogue range. Same signal state is represented within band at all levels. Relatively there are small changes to the analogue signal levels because of manufacturing tolerance, signal attenuation. Generally in most cases there are two states and they are represented by two voltage bands: one is 0 and another is 1. Digital technique is very useful due to its easiness to get an electronic device to switch into one of the number of known states.

The advantages of digital electronics circuit are given below:

1. Signals represented digitally can be transmitted without degradation.
2. More precise representation of a signal can be obtained.
3. Additional resolution is not required for fundamental improvements in the linearity and noise characteristics.
4. Computer-controlled digital systems can be controlled by software.
5. New functions to be added without changing hardware.
6. Easy to store information.

The disadvantages of digital electronics circuit are given below:

1. Digital electronics consumes more energy.
2. Digital electronics produces more heat which increases the complexity of the circuits.
3. More expensive sometimes.
4. Digital fragility can be reduced.
5. Error detection and correction use additional data to correct any errors in transmission and storage.
6. A single-bit error in audio data stored directly as linear pulse code modulation.

1.3.2 Digital Signal Processing (DSP)

The study of signals in a digital representation forms and the processing methods of these signals are called digital signal processing. The analogue signal processing are subsets of signal processing. There are three subfields of digital signal processing, namely, audio signal processing, digital image processing and speech processing.

Digital signal processing is a technique for mathematical manipulation of an information signal. It is defined by the representation of digital time, digital frequency or other digital domain signals by a sequence of symbols or number and the processing of all these signals.

1.4 DESIGN ISSUES IN DIGITAL CIRCUIT

The building blocks of digital circuits are analogue components. The digital circuit is designed with the help of analogue components. At the time of designing the digital circuit, it must be assured that the analogue nature of the components does not dominate the desired digital behaviour. It is also taken into consideration that the digital systems must manage filter power connections, timing margins, parasitic inductances and capacitances, and noises.

Faulty designs of digital circuit have problems such as "glitches", vanishingly fast pulses that may trigger some logic but not others, "runt pulses" that do not reach valid "threshold" voltages, or unexpected ("undecoded") combinations of logic states. (Y.K. Chan and S.Y. Lim, Faculty of Engineering & Technology, Multimedia University, Malaysia.)

Whenever clocked digital systems interface to analogue systems or systems that are driven from a different clock pulses, the digital system can be subject to metastability where a change to the input violates the set-up time for a digital input latch. This situation will be self-resolved, but it will take a random time, and while it persists it can be resulted in invalid signals being propagated within the digital system for a short time.

It is known that digital circuits are designed from analogue components, the digital circuits calculate slower than low-precision analogue circuits that use a similar amount of space and power.

1.5 DIGITAL COMPUTER

Digital computer is one kind of device which is capable of solving problems by processing information in the discrete form. The computer operated on discrete or digital data is known as digital computer. The digital computer is also operated on magnitudes, letters and symbols by representing it in binary form or by converting in digital form. It can also process the business data which is in discrete form. The digital computer is generally used in education and in scientific computation.

1.5.1 Digital Computer Development

French scientist Blaise Pascal and German scientist Gottfried Wilhelm Leibniz developed mechanical digital calculating machines during the 17th century. Charles Babbage, a British inventor developed the first automatic digital computer in the year 1830 called Analytical Engine, which was a mechanical device designed to combine basic arithmetic operations with decisions based on its own computations.

1.5.2 Digital Computer Functional Element

Digital computer has five basic functional elements which are shown in the block diagram of digital computer in Figure 1.2.

4 Digital Logic Design

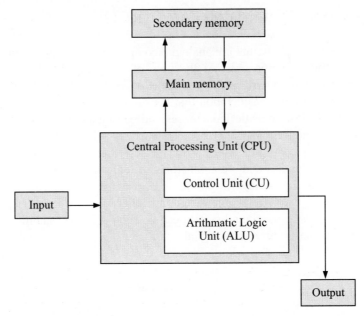

Figure 1.2 Block diagram of digital computer.

1. Input
2. Output
3. Main memory
4. Control unit
5. Arithmatic logic unit (ALU)

Input: This device is used to enter data, programme and instruction into computer. The examples of input devices are keyboard, mouse, scanner, etc.

Output: Devices produce the output after processing operation by the computer. The examples of output devices are printer, CRT monitor, LCD monitor, etc.

Main memory: The device used to store the input data as well the result produced by computer. The example of main memory is RAM.

Control unit: It selects and calls up instructions from the memory in appropriate sequence and relays the proper commands to the appropriate unit for processing.

Arithmatic logic unit (ALU): This unit is used to perform arithmetic as well as logical operation.

1.6 SPECIFICATION AND IMPLEMENTATION OF DIGITAL DESIGN

When a digital system is designed, the specification analysis is done, in which it is investigated that what are the requirements to design a digital system. At the time of designing the digital system, implementation of digital system is taken into consideration, how the designed digital system will be implemented. The compatibility with existing system is also considered at the time of designing the digital system. The specification and implementation of digital design are shown in Figure 1.3. This figure describes the relationship between specification and implementation of digital design. The specifications describe the function of digital system. The digital system becomes the component of a complex system. The component of the complex digital system is also called the module of complex system. One complex system consists with various modules to serve a particular task. The specification is the basis for implementation of a digital system by connecting simpler components.

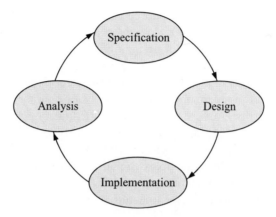

Figure 1.3 Specification and implementation of digital design.

1.6.1 Specification

The design of digital system is always beginning with specification that describes about the digital system which is to be designed. The characteristics and functionality of digital system are described here. The detailed analysis is performed before designing the digital system. The relationship between input and output is also investigated. The Boolean functions are determined with the help of relationship between input and output of the digital system. The various specification levels are: High level, Binary level, Algorithm level; specification of combination circuit is given in Chapter 7 and specification of sequential circuit is given in Chapter 8.

1.6.2 Implementation

The various modules are interconnected to make a digital circuit level network. There are several levels of digital circuit implementation depending on complexity of primitives' module from simple gates to complex processors. There are three ways to implement digital system, which are given below:

- Hierarchical implementation
- Top-down approach
- Bottom-up approach

The digital system is implemented by interconnection of electronic elements such as transistors, resistors and combination of integrated circuit by board. The elements are connected via connector and buses. The various elements are all type of memories like RAM, ROM, SDRAM, etc., microprocessor and peripheral devices. The implementation of combinational system is described in Chapter 7 and sequential system in Chapter 8.

1.7 STRUCTURED AND TRIAL METHODS IN DESIGN

We think for a digital circuit that performs particular task or function. We consider the structure of digital logic before designing the circuit. How the components are connected to each other for producing the desired result? At the time of designing a digital circuit, we focus on a systematic process of circuit design that is based on structure expressed in a hardware description language. The digital design process systematically guides us to develop a complex system which helps us to meet the requirement of modern application. Systematic approach is required when we are designing a digital system of significant complexity. When many persons are collaborating on a design then it becomes specially important. How many people can work to develop a digital system, it depends on the complexity of system. The size of a design team may be from one person to many persons to develop a complex digital system.

The term "structure and trial methods" is used in reference to the systematic process of design, verification and preparation for manufacture of a digital system as a product.

1.8 CAD TOOLS

Digital logic circuit designer uses manual approach to design small digital logic circuit. It is very difficult to design logic circuit for complex system which is used in today's world. Logic circuits found in complex systems as today's computers that cannot be designed manually. The complex circuits are designed using sophisticated CAD tools that automatically implement the synthesis techniques. There are a number of CAD tools which are packaged together into CAD system which is helpful in the following tasks:

1. Design entry
2. Synthesis and optimization
3. Simulation
4. Physical design

CAD systems is very useful for digital logic designer to view a design from different angles with push button and designer can also zoom in or out for close-ups and long-distance views. There are various CAD tools which are available in market for designing digital logic circuit. The detail descriptions about the CAD tools are described in Section 1.8.1. Section 1.8.1 also describes how to get that particular CAD tool for a particular digital logic design. The typical diagram of CAD system is shown in Figure 1.4.

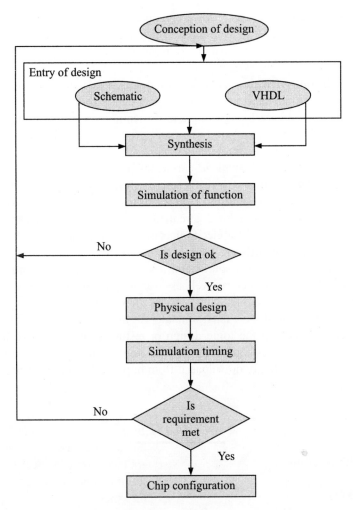

Figure 1.4 CAD system.

1.8.1 CAD Tools Available for Designing Digital Logic Circuit

There are a number of CAD tools which are available in market for designing the digital circuit. Some of the CAD tools are as follows:

- **Xilinx and ModelSim:** This tool is useful for designing digital circuit. You can download and install this tool. This tool is freely available as student version. You can download it from http://www.xilinx.com/support/download.html.
- **Electronic Design Automation (EDA or ECAD):** This software is used for designing electronic systems such as printed circuit boards and integrated circuits. This tool is used to design and analyze all the chips of semiconductor. You can download it from http://www.mccad.com/TotallyFreeMenu.html.

8 Digital Logic Design

- **PCB Design Software:** This software is used to develop your printed circuit board faster and in easy ways. It is available as open source PCB design editor for Microsoft Windows. It is very easy to install. You can download it from http://www.pad2pad.com/_download/p2psetup1998.exe.
- **Circuits Diagram Software:** This software is used to create circuit diagram or digital logic diagram with standard symbol and smart connector. The built-in template is also available in this software for creating and presenting your circuit and logic diagram in minimum time. It can also create circuit and print circuit board diagram, logic designs (digital and analogue) and integrated circuit schematics. You can download it from http://www.edrawsoft.com/download-edrawmax.php.
- **MultiMedia Logic Digital Circuit Design Simulator:** This is multimedia based simulator. It is very simple and most powerful tool. You can design your circuit easily with the help of multimedia logic circuit design simulator. You can download it for 32-bit machine from http://www.electronicecircuits.com/electronic-software/digital-logic-circuit-design-simulator-software.
- **Logic Circuit:** It is an open source and free software for simulating designing digital logic circuit. It is educational as well as user graphic software. You can download it from http://logiccircuit.codeplex.com/releases/view/119633.
- **CEDAR Logic Simulator:** This simulator is an interactive digital logic simulator. This is used for teaching the logic design or testing simple digital designs. Low-level logic gates and high-level component features are available in this simulator. It is also very simple. You can simply download it from http://sourceforge.net/projects/cedarlogic/files/latest/download.

EXERCISES

Short Answer Questions

1. What is CAD tool?
2. List the name of some CAD tools available in market.
3. What are digital computer functional elements?
4. What is digital computer?
5. What are design issues in digital circuit?
6. What is digital electronics?
7. Define term digital.
8. What is digital signal processing (DSP)?

Long Answer Questions

1. Draw the diagram of CAD system.
2. Explain structured and trial methods in design.

3. Why CAD tools play important role in circuit design?
4. What are functional elements of digital computer? Explain all elements of digital computer.
5. Discuss briefly specification and implementation of digital design.
6. Explain various CAD tools available in market for designing a digital circuit.
7. Discuss digital computer development.
8. Draw block diagram of digital computer and explain each unit.

Multiple Choice Questions

1. Who had developed mechanical digital calculating machines?
   (a) Blaise Pascal and Gottfried Wilhelm Leibniz
   (b) William Stallings
   (c) Blaise Pascal only
   (d) Gottfried Wilhelm Leibniz only
2. Which of the following is advantage of digital electronics in compare to analogue circuit?
   (a) Signals represented digitally can be transmitted with degradation
   (b) Signals represented digitally can be transmitted without degradation
   (c) Signals represented digitally can be transmitted with and without degradation
   (d) None
3. The study of signals in a digital representation form and the processing methods of these signals are called:
   (a) Digital signal processing   (b) Digital electronics
   (c) Digital computer   (d) None
4. How many basic functional elements of a digital computer?
   (a) 5   (b) 4
   (c) 3   (d) 6
5. The data are represented in high voltage and low voltage are known as
   (a) Analogue   (b) Continuous
   (c) Digital   (d) Digital and Analogue
6. Which of the following is a CAD tool for digital circuit design?
   (a) Xilinx and ModelSim   (b) Zen
   (c) Xen   (d) C#
7. Which of the following software is used for printed circuit board faster and in easy ways?
   (a) BCP design software   (b) PCB design software
   (c) CPB design software   (d) BPC design software
8. An interactive digital logic simulator is
   (a) CEDAR logic simulator   (b) ECDAR logic simulator
   (c) CEADR logic simulator   (d) CEDRA logic simulator
9. Which of the following is the way to implement digital system?
   (a) Hierarchical implementation   (b) Linear implementation
   (c) Hybrid implementation   (d) Hybrid hierarchical implementation

10. One complex system consists with
    (a) A module
    (b) Two modules
    (c) Various modules
    (d) (a) and (b)

Answers

1. (a)	**2.** (b)	**3.** (a)	**4.** (a)	**5.** (c)	**6.** (a)
7. (b)	**8.** (a)	**9.** (a)	**10.** (c)		

BIBLIOGRAPHY

Ashenden, P.J., *Digital Design: An Embedded Systems Approach Using Verilog*, Morgan Kaufmann Publishers, 2008.

Dally, W.J. and Harting, R.C., *Digital Design: A Systems Approach*, Cambridge University Press, 2012.

https://inst.eecs.berkeley.edu/~ee100/su07/handouts/IntroductionToDigitalSystems.pdf .

http://www.britannica.com/EBchecked/topic/163278/digital-computer.

http://www.webopedia.com/TERM/C/CAD.html.

Chapter 2

Number Systems

2.1 INTRODUCTION

The number systems are very useful to the people who work with computing. We are familiar with decimal number only which is generally used day to day life. But computer cannot understand the decimal number. Computer can understand 0 (low voltage) and 1 (high voltage) only, so there is a communication gap between human being and computer. A system is introduced to fill the communication gap and this gap is number system. The number system is a group of various type of number, which is interchangeable, means one type of number can be converted to another type of number and vice versa.

2.2 DEFINITION

The number system is defined as a set of numerals which are used to for representing the numbers. The algebraic definition of number system is a set of numbers used to solve different type of algebra problems. The number systems in algebra are integer numbers, real numbers, rational numbers, natural numbers, irrational number, odd number and even number, etc. There are four number system generally used in computer, namely, binary number system, decimal number system, octal number system and hexadecimal system.

2.3 BINARY NUMBER

The binary number is as set of two digits zero (0) and one (1). Binary number is define as a set of 0 and 1, let B is a binary number then mathematically it can be represented as $B = \{0, 1\}$. There are two elements in set B so the base of binary number is two. The base is also known as radix on number system. The counting numbers of binary systems are given in Table 2.1.

Table 2.1 Binary counting number

0	1	10	11	100	101	110	111
1000	1001	1010	1011	1100	1101	1110	1111
10000	10001	10010	10011	10100	10101	10110	10111
11000	11001	11010	11011	11100	11101	11110	11111

Put 1 in front of numbers 0 to 1 to obtain counting numbers from 10 to 11, and then put 10 in front of numbers 0 to 1, through this method you will obtain counting numbers from 100 to 101. Similarly, you can obtain counting number corresponding to all possible binary numbers directly.

2.4 OCTAL NUMBER

The octal number is a set of seven digits 0, 1, 2, 3, 4, 5, 6 and 7. Let O is an octal number then mathematically it can be represented as $O = \{0, 1, 2, 3, 4, 5, 6, 7\}$. There are eight elements in set O, therefore the base of octal number is eight. The only 0 to 7 numbers are involved in an octal number. The counting numbers of octal numbers are given in Table 2.2.

Table 2.2 Octagonal counting number

0	1	2	3	4	5	6	7
10	11	12	13	14	15	16	17
20	21	22	23	24	25	26	27
30	31	32	33	34	35	36	37

Put 1 in front of numbers 0 to 7 to obtain counting numbers from 10 to 17, put 2 in front of numbers 0 to 7 to obtain counting numbers from 20 to 27 similarly you can put 3, 4, 5, in front of numbers from 0 to 7 to obtain all possible octal numbers directly.

2.5 HEXADECIMAL NUMBER

The hexadecimal number is a set of fifteen digits 0, 1, 2, 3, 4, 5, 6, 7, 8, 9 A, B, C, D, E and F. Let H is a hexadecimal number then mathematically it can be represented as $H = \{0, 1, 2, 3, 4, 5, 6, 7, A, B, C, D, E, F\}$. There are fifteen elements in set H; therefore the base of hexadecimal number is sixteen. Only 0 to F numbers are involved in a hexadecimal number. The counting numbers of hexadecimal numbers are given in Table 2.3.

Table 2.3 Hexagonal counting number

0	1	2	3	4	5	6	7	8	9	*A*	*B*	*C*	*D*	*E*	*F*
10	11	12	13	14	15	16	17	18	19	1A	1B	1C	1D	1E	1F
20	21	22	23	24	25	26	27	28	29	2A	2B	2C	2D	2E	2F
30	31	32	33	34	35	36	37	38	39	3A	3B	3C	3D	3E	3F

Put 1 in front of numbers 0 to F to obtain counting numbers from 10 to 1F, put 2 in front of numbers 0 to F to obtain counting numbers from 20 to 2F similarly you can put 3, 4, 5, in front of numbers from 0 to F to obtain all possible hexadecimal numbers directly.

2.6 HEXAL NUMBER

The hexal number is a set of six digits 0, 1, 2, 3, 4 and 5. Let M is a hexal number then mathematically it can be represented as $M = \{0, 1, 2, 3, 4, 5\}$. There are six elements in set M; therefore, the base of hexadecimal number is six. Only 0 to 5 numbers are involved in a hexal number. The counting numbers of hexal numbers are given in Table 2.4.

Table 2.4 Hexal counting number

0	1	2	3	4	5
10	11	12	13	14	15
20	21	22	23	24	25
30	31	32	33	34	35

Put 1 in front of numbers 0 to 5 to obtain counting numbers from 10 to 15, put 2 in front of numbers 0 to 5 to obtain counting numbers from 20 to 25 similarly you can put 3, 4, 5, in front of numbers from 0 to 5 to obtain all possible hexal numbers directly.

2.7 NANO NUMBER

The nano number is a set of nine digits 0, 1, 2, 3, 4, 5, 6, 7 and 8. Let N is a nano number then mathematically it can be represented as $N = \{0, 1, 2, 3, 4, 5, 6, 7, 8\}$. There are nine elements in set N; therefore the base of nano number is nine. Only 0 to 8 numbers are involved in a nano number. The counting numbers of nano numbers are given in Table 2.5.

Table 2.5 Nano counting number

0	1	2	3	4	5	6	7	8
10	11	12	13	14	15	16	17	18
20	21	22	23	24	25	26	27	28
30	31	32	33	34	35	36	37	38

Put 1 in front of numbers 0 to 8 to obtain counting numbers from 10 to 18, put 2 in front of numbers 0 to 8 to obtain counting numbers from 20 to 28 similarly you can put 3, 4, 5, in front of numbers from 0 to 8 to obtain all possible nano numbers directly.

2.8 DECIMAL TO BINARY CONVERSION

The changing base of 10 of a number to base of 2 is known as decimal to binary conversion. Divide the given decimal number with the base of binary number(2) and note the remainder, finally write the noted remainder in reverse order, the resultant is binary equivalent of a given decimal number. The examples are given below:

EXAMPLE 2.1 Convert decimal number 24 into binary number.

Solution:

Divisor	Number	Remainder
2	24	Initial
2	12	0 (LSB)
2	6	0
2	3	0
2	1	1
2	0	1 (MSB)

The remainder is 00011, the reverse order of remainder is 11000, and so binary number 11000 is equivalent to decimal number 24 or writes remainder from MSB to LSB.

EXAMPLE 2.2 Convert decimal number 25 to binary number.

Solution:

Divisor	Number	Remainder
2	25	Initial
2	12	1 (LSB)
2	6	0
2	3	0
2	1	1
2	0	1 (MSB)

The remainder is 10011, the reverse order of remainder is 11001, and so binary number 11001 is equivalent to decimal number 24 or writes remainder from MSB to LSB.

2.8.1 The Conversional of Floating Point Decimal Number to Binary Number

This decimal number has two parts: mantissa part and exponent part. The number before decimal is mantissa part and number after decimal is exponent part. To obtain the binary equivalent to the above given number by dividing continuously the mantissa part with the base(2) of binary number and exponent part can be multiplied by the base(2) of binary number and recording the number before the decimal point, do it continuously until you get the zero result. The recoded number is your binary number of exponent part. Place together the number by separating mantissa part and exponent part with the decimal point.

Number Systems

EXAMPLE 2.3 Convert decimal number to 0.20 to binary number.

Solution: The decimal number is followed by period (decimal point), in this case multiply the number by 2

$$0.20 \times 2 = 0.40$$
$$0.40 \times 2 = 0.80$$
$$0.80 \times 2 = 1.60$$
$$0.60 \times 2 = 1.20$$

In the last step of the above example, the number is repeating. As the number is repeating stop the multiplication and note down the number before the decimal point in the right-hand side from top to bottom. The result obtained in the above example is 0.0011.

EXAMPLE 2.4 Convert the decimal number 0.25 to binary number.

Solution:
$$0.25 \times 2 = 0.50$$
$$0.50 \times 2 = 1.0$$

In the last step of the above example, the number after decimal point in the right-hand side becomes zero, and then you can stop and note the number before decimal point from top to bottom in the right-hand side. The result of the above example is 0.01.

EXAMPLE 2.5 Convert the number 0.23 to binary number

Solution:
$$0.23 \times 2 = 0.46$$
$$0.46 \times 2 = 0.92$$
$$0.92 \times 2 = 1.82$$
$$0.82 \times 2 = 1.64$$

$$0.64 \times 2 = 1.28$$
$$0.28 \times 2 = 0.56$$
$$0.56 \times 2 = 1.12$$

In such type of example neither number repeats nor zero, in this case do the multiplication up to eight or nine iteration or number of iteration is given in the question. The result of the above example is 0.00110101.

EXAMPLE 2.6 Convert the number 25.20 to binary number.

Solution: This example is a combination of mantissa and exponent parts. Divide the mantissa part by 2 as in Example 2.1 and multiply by 2 as in Example 2.4 separately. After getting the results separately, combine both the results together, separated by decimal point. The binary conversion of mantissa part (25) is as follows:

Divisor	Number	Remainder
2	25	Initial
2	12	1 (LSB)
2	6	0
2	3	0
2	1	1
2	0	1 (MSB)

16 Digital Logic Design

Result: 11001

The binary conversion of exponent part (0.20) is as follows:

$$0.20 \times 2 = 0.40$$
$$0.40 \times 2 = 0.80$$
$$0.80 \times 2 = 1.60$$
$$0.60 \times 2 = 1.20$$

Result: 0.0011

The combination of both the results is 11001.0011; therefore, the binary equivalent of decimal number 25.20 is 11001.0011.

2.8.2 Direct Method

This method is used to directly convert the decimal number to binary number without division. Write number 1, 2, 4, 8, 16, 32, 64, from left to right. For the conversion of decimal number to binary number, put one in front of above mentioned sequence and find the sum, if the sum is equivalent to desired number which is to be converted and place zero in place of vacant place in the above sequence.

EXAMPLE 2.7 Convert the decimal number 48 to binary number directly.

Solution:

—	64	32	16	8	4	2	1
—	—	1	1	0	0	0	0

32 + 16 = 48, so place 1 in front of 32 and 16, put 0 in place of vacant place from right to left from 32, and note down the combination of 0 and 1 in the above table, which will be the result. The binary equivalent of decimal number 48 is 110000.

2.9 BINARY TO DECIMAL CONVERSION

Binary to decimal conversion means changing the base 2 of binary number to base 10 of decimal number. Count the total digit in the given binary number, multiply first digit by 2 to the power of total digit minus 1, add it to next digit then multiply by 2 to the power of total digit minus 2, do this continuously till you reach at the last digit of binary number. Let X is the binary number and its digits are $a_0, a_1, a_2, a_3, a_4, \cdots, a_n$, and Y is the decimal number equivalent of binary number X. Then the general formula for converting binary number to decimal number is $a_0 \times 2^n + a_1 \times 2^{n-1} + a_2 \times 2^{n-2} + a_3 \times 2^{n-3} + a_4 \times 2^{n-4} + \cdots + a_n \times 2^{n-i}$, where $i = 0, 1, 2, \cdots, n$. It can be also represented in short form as decimal number $\sum_{i=0}^{n} a_i \times 2^{n-i}$.

EXAMPLE 2.8 Convert the binary number 1100101 to the decimal number.

Solution: The total number in the binary number 1101101 is 7. Perform the operations as follows:

$$1 \times 2^{7-1} + 1 \times 2^{7-2} + 0 \times 2^{7-3} + 1 \times 2^{7-4} + 1 \times 2^{7-5} + 0 \times 2^{7-6} + 1 \times 2^{7-7}$$

Simplification

$1 \times 2^6 + 1 \times 2^5 + 0 \times 2^4 + 1 \times 2^3 + 1 \times 2^2 + 0 \times 2^1 + 1 \times 2^0$

$1 \times 64 + 1 \times 32 + 0 \times 16 + 1 \times 8 + 1 \times 4 + 0 \times 2 + 1 \times 1$ (where $2^0 = 0$)

$64 + 32 + 0 + 8 + 4 + 0 + 1 = 109$

The decimal number 109 is equivalent of binary number 1101101.

2.9.1 Binary Fraction Number to Decimal Conversion

Multiply each digit of binary fraction number by 2 to the power minus 1 and add it to next digit then multiply by 2 to the power minus 2, do it until the end of fraction digit. Let X is the binary fraction number and its digits are $a_0, a_1, a_2, a_3, a_4, \cdots, a_n$, and Y is the decimal number equivalent to binary fraction number X. The general formula for binary fraction to decimal conversion is $a_1 \times 2^{-1} + a_2 \times 2^{-2} + a_3 \times 2^{-3} + a_4 \times 2^{-4} + a_5 \times 2^{-5} + a_n \times 2^{-n}$, where $n = 1, 2, \cdots, n$. It can be also represented in short form as decimal number $Y = \sum_{i=1}^{n} a_i \times 2^{-n}$.

EXAMPLE 2.9 Convert binary fraction 0.110101 to decimal number.

Solution: $1 \times 2^{-1} + 1 \times 2^{-2} + 0 \times 2^{-3} + 1 \times 2^{-4} + 0 \times 2^{-5} + 1 \times 2^{-6}$

Simplification

$1 \times 1/2 + 1 \times 1/2^2 + 0 \times 1/2^3 + 1 \times 1/2^4 + 0 \times 1/2^5 + 1 \times 1/2^6$

$1 \times 1/2 + 1 \times 1/4 + 0 \times 1/8 + 1 \times 1/16 + 0 \times 1/32 + 1 \times 1/64$

$1 \times 0.5 + 1 \times 0.25 + 0 \times 0.125 + 1 \times 0.0625 + 0 \times 0.03125 + 1 \times 0.015625$

$0.5 + 0.25 + 0 + 0.0625 + 0 + 0.015625$

0.8228125, therefore $(0.110101)_2 = (0.8228125)_{10}$.

EXAMPLE 2.10 Convert binary number 11101.1101 to decimal number.

Solution: In this example, the binary number is consisting with non-fraction and fraction numbers. The binary number 11101 is non-fraction number and 0.1101 is a fraction number. Convert both numbers separately and at last combine both converted numbers together to get the result.

Non-fraction Number: 11101

$1 \times 2^4 + 1 \times 2^3 + 1 \times 2^2 + 0 \times 2^1 + 1 \times 2^0$

$1 \times 16 + 1 \times 8 + 1 \times 4 + 0 \times 2 + 1 \times 1$

$16 + 8 + 4 + 0 + 1$

29

The Fraction Part: 0.1101

$1 \times 2^{-1} + 1 \times 2^{-2} + 0 \times 2^{-3} + 1 \times 2^{-4}$

$1 \times 1/2 + 1 \times 1/2^2 + 0 \times 1/2^3 + 1 \times 1/2^4$

$1 \times 1/2 + 1 \times 1/4 + 0 \times 1/8 + 1 \times 1/16$

$1 \times 0.5 + 1 \times 0.25 + 0 \times 0.125 + 1 \times 0.0625$
$0.5 + 0.25 + 0 + 0.0625$
0.8125

Now combine both non-fraction and fraction parts together: 29.8125.
The result is 29.8125 in decimal number of binary number 11101.1101.

2.10 DECIMAL TO OCTAL CONVERSION

The conversion of decimal to octal number means changing the base 10 of decimal to base 8 of octal number. Divide the decimal number by 8 which is the base of octal number and record the remainder, do it until the number is zero. Write the remainder from bottom to top, you will get the result.

EXAMPLE 2.11 Convert the decimal number 85 to octal number.

Solution:

Divisor	Number	Remainder
8	85	Initial
8	10	5 (LSB)
8	1	2
8	0	1 (MSB)

The remainder in the above table is 521; the reverse of the remainder is 125, which is the octal equivalent of decimal number 85.

EXAMPLE 2.12 Convert the decimal number 480 to octal number.

Solution:

Divisor	Number	Remainder
8	480	Initial
8	60	0 (LSB)
8	7	4
8	0	7 (MSB)

The remainder in the above table is 047; the reverse of the remainder is 740, which is the octal equivalent of decimal number 480.

2.10.1 Decimal Fraction to Octal Conversion

Multiply the fraction decimal number by eight (8), and record the number before the decimal point, do this until the number is zero or number starts repeating of up to given iteration or you can assume eight or nine iteration. The recorded number is the result.

EXAMPLE 2.13 Convert the decimal number 0.98 to octal number.

Solution:

Number	Result
$0.98 \times 8 = 7.84$	7
$0.84 \times 8 = 6.72$	6
$0.72 \times 8 = 5.76$	5
$0.76 \times 8 = 6.08$	6
$0.08 \times 8 = 0.64$	0
$0.64 \times 8 = 5.12$	5
$0.12 \times 8 = 0.96$	0
$0.96 \times 8 = 7.68$	7
$0.68 \times 8 = 5.44$	5

In this type of example the number is neither repeated nor zero, so you can stop after eight or nine iteration. The result of this example is 0.765605075.

EXAMPLE 2.14 Convert the decimal number 0.85 to octal number.

Solution:

Number	Result
$0.85 \times 8 = 6.8$	6
$0.8 \times 8 = 6.4$	6
$0.4 \times 8 = 3.2$	3
$0.2 \times 8 = 1.6$	1
$0.6 \times 8 = 4.8$	4
$0.8 \times 8 = 6.4$	Repeat

In this example the number is repeating from 8 so you can stop and write the result from top to bottom without including the number repeating at 8. The result is 0.66314.

EXAMPLE 2.15 Convert the decimal number 85.85 to octal number.

Solution: In this example the fraction part and non-fraction part are combined. You can separately convert the fraction part and non-fraction part. After separately conversion of both parts you combine both together to get the result.

Non-fraction part: 85, the conversion of this number is as follows:

Divisor	Number	Remainder
8	85	Initial
8	10	5 (LSB)
8	1	2
8	0	1 (MSB)

The remainder in the above table is 521; the reverse of the remainder is 125, which is the octal equivalent of decimal number 85.

Fraction part: .85, the conversion of this number is as follows:

Number	Result
0.85 × 8 = 6.8	6
0.8 × 8 = 6.4	6
0.4 × 8 = 3.2	3
0.2 × 8 = 1.6	1
0.6 × 8 = 4.8	4
0.8 × 8 = 6.4	Repeat

In this example the number is repeating from 8 so you can stop and write the result from top to bottom without including the number repeating at 8. The result is 0.66314. Now, you can combine both parts together as 125.66314, this is octal number of decimal number 85.85.

EXAMPLE 2.16 Convert the decimal number 0.25 to octal number.

Solution:

Number	Result
0.25 × 8 = 2.0	2

After one iteration number becomes zero, you cannot move further therefore 0.2 is octal number equivalent to decimal number 0.25.

2.11 OCTAL TO DECIMAL CONVERSION

Octal to decimal conversion means changing the base 8 of octal to the base 10 of decimal. For the conversion from octal to decimal, first count the total number of digits involved in octal number. Multiply first digit of octal number by 8 to the power of total number of digits minus 1 and add it to the next digit then multiply by 8 to the power of total minus 2, and do it until the last digit of octal number, the summation of all digits is the result. Let X is an octal number consisting of $a_0, a_1, a_2, a_3, a_4, \cdots, a_n$, and Y is a decimal equivalent of octal number X. The general formula for converting X into Y is as follows:

$Y = a_0 \times 8^n + a_1 \times 8^{n-1} + a_2 \times 8^{n-2} + a_3 \times 8^{n-3} + a_4 \times 8^{n-4} + \cdots + a_n \times 8^{n-n}$, this can also be written in short for as $Y = \sum_{i=0}^{n} a_i \times 8^{n-i}$.

EXAMPLE 2.17 Convert the octal number 256 to decimal number.

Solution: $2 \times 8^2 + 5 \times 8^1 + 6 \times 8^0$
$2 \times 64 + 5 \times 8 + 6 \times 1$
$128 + 40 + 6$
174

Result is $(174)_{10}$

EXAMPLE 2.18 Convert the octal number 7056 to decimal number.

Solution: $7 \times 8^3 + 0 \times 8^2 + 5 \times 8^1 + 6 \times 8^0$
$7 \times 512 + 0 \times 64 + 5 \times 8 + 6 \times 1$
$3584 + 0 + 40 + 6$
3630

Result is $(3630)_{10}$.

2.11.1 Octal Fraction to Decimal Conversion

Multiply each digit of octal fraction number by 8 to the power minus 1, and add it to the next digit then multiply by 8 to the power minus 2 continuously, do it until the end of fraction digit. Let X is the binary fraction number and its digits are $a_0, a_1, a_2, a_3, a_4, \cdots, a_n$, and Y is decimal number equivalent to binary fraction number X. The general formula for binary fraction to decimal conversion is $a_1 \times 8^{-1} + a_2 \times 8^{-2} + a_3 \times 8^{-3} + a_4 \times 8^{-4} + a_5 \times 8^{-5} + \cdots + a_n \times 8^{-n}$, where $n = 1, 2, \cdots, n$. It can be also represented in short form as decimal number $Y = \sum_{i=1}^{n} a_i \times 8^{-n}$.

EXAMPLE 2.19 Convert the octal number 0.258 to decimal number.

Solution: $2 \times 8^{-1} + 5 \times 8^{-2} + 8 \times 8^{-3}$
$2 \times 1/8 + 5 \times 1/8^2 + 8 \times 1/8^3$
$2 \times 1/8 + 5 \times 1/64 + 8 \times 1/512$
$2 \times 0.125 + 5.00156 + 8.001953$
$0.25 + 0.078 + 0.0156$
0.3436

EXAMPLE 2.20 Convert the octal number 254.258 to decimal number.

Solution: This number is a combination of fraction and non-fraction parts of octal number. In this case, fraction number and non-fraction number are converted separately and then combine it to get the result.

The non-fraction number is 254, so this is converted separately as follows:

$2 \times 8^2 + 5 \times 8^1 + 4 \times 8^0$
$2 \times 64 + 5 \times 8 + 4 \times 1 \quad$ where $8^0 = 1$
$128 + 40 + 4$
172

The fraction number is 0.258, so this is converted as follows:

$2 \times 8^{-1} + 5 \times 8^{-2} + 8 \times 8^{-3}$
$2 \times 1/8 + 5 \times 1/8^2 + 8 \times 1/8^3$
$2 \times 1/8 + 5 \times 64 + 8 \times 1/512$
$2 \times 0.125 + 5 \times 0.0156 + 8 \times 0.001953$
$0.25 + 0.078 + 0.0156$
0.3436

22 Digital Logic Design

Now both combined number is 172.3436, which is decimal number equivalent to octal number 254.258

The result is 172.3436.

2.12 OCTAL TO BINARY CONVERSION

The highest element of octal number is seven (7). The minimum three bits required to represent the 7 in binary. Each digit of octal number is represented in three bit in binary. Each digit of octal number represented binary number all together is the octal conversion to binary number.

EXAMPLE 2.21 Convert octal number $(234)_8$ to binary number.

Solution: The individual digit of octal number 234 is 2, 3 and 4. Each digit in three-bit representation is as follows:

Three-bit binary representation of 2 is

Weight	W3 = 4	W2 = 2	W1 = 1
Binary	B1 = 0	B2 = 1	B3 = 0

Total weight = W3 × B1 + W2 × B2 + W1 × B3
Total weight = 4 × 0 + 2 × 1 + 1 × 0 = 0 + 2 + 0 = 2, so binary representation of 2 is **010**, similarly
  Binary representation of 3 is 4 × 0 + 2 × 1 + 1 × 1 = 3 = **011**
  Binary representation of 4 is 4 × 1 + 2 × 0 + 1 × 0 = 4 = **100**
  The binary representation of octal number $(234)_8$ is **010011100**

OR

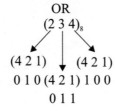

Result is **010 011 100**

2.12.1 Octal Fraction to Binary Conversion

The conversion of octal fraction number to binary fraction number is done by representing each digit of octal fraction number into the three-bit representation and all digit representation combined together and put decimal point as prefix to get the result.

EXAMPLE 2.22 Convert octal number $(0.234)_8$ to binary number.

Solution: Three-bit binary representation of 2 is

Weight	W3 = 4	W2 = 2	W1 = 1
Binary	B1 = 0	B2 = 1	B3 = 0

Total weight = W3 × B1 + W2 × B2 + W1 × B3

Total weight = 4 × 0 + 2 × 1 + 1 × 0 = 0 + 2 + 0 = 2, so binary representation of 2 is **010**, similarly

Binary representation of 3 is 4 × 0 + 2 × 1 + 1 × 1 = 3 = **011**

Binary representation of 4 is 4 × 1 + 2 × 0 + 1 × 0 = 4 = **100**

The binary representation of octal number $(0.234)_8$ is **0.010011100**

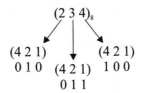

Result is **0.010 011 100**

EXAMPLE 2.23 Convert octal number $(234.23)_8$ to binary number.

Solution: The conversion of $(234)_8$ octal number to binary is as follows:
Three-bit binary representation of 2 is

Weight	W3 = 4	W2 = 2	W1 = 1
Binary	B1 = 0	B2 = 1	B3 = 0

Total weight = W3 × B1 + W2 × B2 + W1 × B3

Total weight = 4 × 0 + 2 × 1 + 1 × 0 = 0 + 2 + 0 = 2, so binary representation of 2 is **010**, similarly

Binary representation of 3 is 4 × 0 + 2 × 1 + 1 × 1 = 3 = **011**

Binary representation of 4 is 4 × 1 + 2 × 0 + 1 × 0 = 4 = **100**

The binary representation of octal number $(234)_8$ is **010011100**.
Similarly,
The conversion of $(0.23)_8$ octal number to binary is as follows
Three-bit binary representation of 2 is

Weight	W3 = 4	W2 = 2	W1 = 1
Binary	B1 = 0	B2 = 1	B3 = 0

Total weight = W3 × B1 + W2 × B2 + W1 × B3

Total weight = 4 × 0 + 2 × 1 + 1 × 0 = 0 + 2 + 0 = 2, so binary representation of 2 is 010, similarly

Binary representation of 3 is 4 × 0 + 2 × 1 + 1 × 1 = 3 = 011.

The combined binary representation of digits 2 and 3 is 010 011, put decimal point before combined number, and then the result is 0.010011.

Now, combined non-octal fraction conversion and fraction octal conversion together is the result for conversion of $(234.23)_8$ to binary number. Result is **010011100.010011**.

2.13 BINARY TO OCTAL CONVERSION

Binary to octal conversion is done by making group of three bits from left to right and convert each group to decimal number and combine all together to get the octal equivalent to the given binary number. Here we are making group three bits only because minimum bits required for representing the highest element (7) in octal number. If you are making a group from left to right, the final group having only two or one digit, put one zero or two zeros to make three digit before the group. Putting zero before last group will not affect the number.

EXAMPLE 2.24 Convert binary number $(11011011)_2$ to octal number.

Solution: The three digits from left to right is 011 (Group-1)

The three digits from Group-1 is 110 (Group-2)

The three digits from Group-2 is 11, here is only two digits but you have to make a group of three, so put zero before two digits 11 like 011, this is Group-3. Now convert each group in decimal number by adding the weight of each digit in a group, Group-1 = 011, the weight from right to left is 1, 2, 4; multiply first digit by 1 and second digit by 2 and third digit by 4 and the sum of these all are the value of Group-1, find the value of Group-2 and Group-3, the result is Group-3Group-2Group-1.

The value of Group-1 is $4 \times 0 + 2 \times 1 + 1 \times 1 = 0 + 2 + 1 = 3$
The value of Group-2 is $4 \times 1 + 2 \times 1 + 1 \times 0 = 4 + 2 + 0 = 6$
The value of Group-3 is $4 \times 0 + 2 \times 1 + 1 \times 1 = 0 + 2 + 1 = 3$
The result is Group-3Group-2Group-1 = $(363)_8$.

Therefore $(363)_8$ is octal number equivalent to binary number $(11011011)_2$
Result is $(363)_8$.

2.13.1 Binary Fraction to Octal Conversion

The binary fraction to octal conversion is done by making group of three binary digits from right to left. Convert each group to decimal number and after conversion combine all groups together to get the result. If in the last group number digit is less than three digits, you add zero to the last digit of group to make three digits in a group. Putting zero to the last digit of group will not affect the number.

EXAMPLE 2.25 Convert binary number $(0.11011011)_2$ to octal number.

Solution: Making group of three digits from right to left as follows:

Group-1 = 110
Group-2 = 110
Group-3 = 11, here is only two digits so add zero after the last digit to make the three digits in this group, Group-3 = 110. Now, find the decimal value of each group:

The value of Group-1 = 110 = $4 \times 1 + 2 \times 1 + 1 \times 0 = 4 + 2 + 0 = 6$
The value of Group-2 = 110 = $4 \times 1 + 2 \times 1 + 1 \times 0 = 4 + 2 + 0 = 6$
The value of Group-3 = 110 = $4 \times 1 + 2 \times 1 + 1 \times 0 = 4 + 2 + 0 = 6$

All the groups combine together are Group-1Group-2Group-3 = 0.666, so the octal number $(0.666)_8$ is equivalent to binary number $(0.11011011)_2$.

EXAMPLE 2.26 Convert binary number $(11101.1101)_2$ to octal number.

Solution: In this example fraction and non-fraction binary numbers are combined. Both numbers are converted separately and the result is combined together.

Non-fraction binary number = 11101

Make the group of three digits from left to right. Group-1 = 101, Group-2 = 11, in Group-2 only two digits are involved, so put zero before the first digit to make a group of three like Group-2 = 011.

The value of Group-1 = 101 = $4 \times 1 + 2 \times 0 + 1 \times 1 = 4 + 0 + 1 = 5$
The value of Group-2 = 011 = $4 \times 0 + 2 \times 1 + 1 \times 1 = 0 + 2 + 1 = 3$

All groups together are Group-2Group-1 = 35, it is the octal number equivalent to binary number 11101.

Fraction binary number = 0.1101

In this number make group of three digits from right to left. Group-1 = 110, Group-2 = 1, Group-2 has only one digit, so put two zeros after the digit to make three digits in this group like Group-2 = 100.

The value of Group-1 = 110 = $4 \times 1 + 2 \times 1 + 1 \times 0 = 4 + 2 + 0 = 6$
The value of Group-2 = 100 = $4 \times 1 + 2 \times 0 + 1 \times 0 = 4 + 0 + 0 = 4$

All groups combined together are Group-1Group-2 = 0.64, this octal number is equivalent to binary number 0.1101. Now, put both faction number and non-fraction number together to get the result. Both the combined numbers are $(35.64)_8$, this octal number is equivalent to binary number $(11101.1101)_2$.

2.14 DECIMAL TO HEXADECIMAL CONVERSION

The conversion from decimal to hexadecimal is done by dividing the given decimal number by base –15 of hexadecimal and record the remainder, do it until number is zero. The decimal number equivalent to given hexadecimal number is reverse of remainder.

EXAMPLE 2.27 Convert decimal number $(548)_{10}$ to hexadecimal number.
Solution:

Divisor	Number	Remainder
16	548	Initial
16	34	4 (LSB)
16	2	2
16	0	2 (MSB)

The remainder is 422. The reverse of remainder is 224, so this hexadecimal number equivalent to decimal number $(548)_{10}$, the result is $(224)_{16}$.

2.14.1 Decimal Fraction to Hexadecimal Conversion

Multiply the decimal fraction number by base 16 of hexadecimal number and record the digit before decimal point in the multiplication result and again multiply after the decimal point number, do it until number is zero or start repeating of neither repeating nor zero, do it up to six or seven or number of iteration given in the question and stop. Record the result from top to bottom.

EXAMPLE 2.28   Convert decimal number $(0.55)_{10}$ to hexadecimal number.

Solution:

Number	Result
$0.55 \times 16 = 8.8$	8
$0.8 \times 16 = 12.8$	C
$0.8 \times 16 = 12.8$	Repeat

After 0.8 number starts repeating so you can stop multiplication. The result is $(8C)_{16}$.

EXAMPLE 2.29   Convert decimal number $(0.58)_{10}$ to hexadecimal number.

Solution:

Number	Result
$0.58 \times 16 = 9.28$	9
$0.28 \times 16 = 7.68$	7
$0.68 \times 16 = 10.88$	A
$0.88 \times 16 = 14.08$	E
$0.08 \times 16 = 1.28$	1
$0.28 \times 16 = 7.68$	Repeat

After 0.28 number starts repeating, so you can stop multiplication. The result is $(97AE1)_{16}$.

EXAMPLE 2.30   Convert decimal number $(0.75)_{10}$ to hexadecimal number.

Solution:

Number	Result
$0.75 \times 16 = 12.0$	C
$0.0 \times 16 = 0$	Number = 0

Hereafter first iteration the number after decimal point is zero, so you can stop.

2.15 HEXADECIMAL TO DECIMAL CONVERSION

Hexadecimal to decimal conversion means changing the base 16 of hexadecimal number to the base 10 of decimal. For the conversion from hexadecimal to decimal, first count the total number of digits involves in hexadecimal number. Multiply first digit of hexadecimal number by 16 to the power of total number of digit minus 1 and add it to the next digit then multiply by 16 to the power of total number minus 2, and continuously do it until the last digit of hexadecimal number, the summation of all digits is the result. Let X is a hexadecimal number consisting of $a_0, a_1, a_2, a_3, a_4, \cdots, a_n$, and Y is a decimal equivalent of hexadecimal number X. The general formula for converting X into Y is as follows:

$Y = a_0 \times 16^n + a_1 \times 16^{n-1} + a_2 \times 16^{n-2} + a_3 \times 16^{n-3} + a_4 \times 16^{n-4} + \cdots + a_n \times 16^{n-n}$, this can also be written in short for as $Y = \sum_{i=0}^{n} a_1 \times 16^{n-i}$.

EXAMPLE 2.31 Convert the hexadecimal number $(256)_{16}$ to decimal number.
Solution: $2 \times 16^2 + 5 \times 16^1 + 6 \times 16^0$
$2 \times 256 + 5 \times 16 + 6 \times 1$
$512 + 80 + 6$
598
The result is $(598)_{10}$.

EXAMPLE 2.32 Convert the hexadecimal number $(2A6C)_{16}$ to decimal number.
Solution: $2 \times 16^3 + A \times 16^2 + 6 \times 16^1 + C \times 16^0$
$2 \times 4096 + 10 \times 256 + 6 \times 16 + 12 \times 1$ (where A = 10 and C = 12)
$8192 + 2560 + 96 + 12$
10860
The result is $(10860)_{10}$.

2.15.1 Hexadecimal Fraction to Decimal Conversion

Multiply each digit of hexadecimal fraction number by 16 to the power minus 1 and adds it to next digit then multiply by 16 to the power minus 2, continuously do it until the end of fraction digit. Let X is the hexadecimal fraction number and its digits are $a_0, a_1, a_2, a_3, a_4, \cdots, a_n$, and Y is the decimal number equivalent to hexadecimal fraction number X. The general formula for hexadecimal fraction to decimal conversion is $a_1 \times 16^{-1} + a_2 \times 16^{-2} + a_3 \times 16^{-3} + a_4 \times 16^{-4} + a_5 \times 16^{-5} + \cdots + a_n \times 16^{-n}$, where $n = 1, 2, \cdots, n$. It can be also represented in short form as decimal number $Y = \sum_{i=1}^{n} a_i \times 16^{-n}$.

EXAMPLE 2.33 Convert $(0.258)_{16}$ hexadecimal number to decimal number.
Solution: $2 \times 16^{-1} + 5 \times 16^{-2} + 8 \times 16^{-3}$
$2 \times 1/16 + 5 \times 1/16^2 + 8 \times 1/16^3$
$2 \times 1/16 + 5 \times 1/256 + 8 \times 1/4096$

$2 \times 0.0625 + 5 \times 0.003906 + 8 \times 0.000244$
$0.125 + 0.01953 + 0.001952$
0.146482

The result is $(0.146482)_{10}$.

EXAMPLE 2.34 Convert $(0.2AD)_{16}$ hexadecimal number to decimal number.

Solution: $2 \times 16^{-1} + A \times 16^{-2} + D \times 16^{-3}$
$2 \times 1/16 + 10 \times 1/16^2 + 13 \times 1/16^3$ (where A = 10 and D = 13)
$2 \times 1/16 + 10 \times 1/256 + 13 \times 1/4096$
$2 \times 0.0625 + 10 \times 0.003906 + 13 \times 0.000244$
$0.125 + 0.03906 + 0.03172$
0.54732

The result is $(0.54732)_{10}$.

EXAMPLE 2.35 Convert $(2A6C.2AD)_{16}$ hexadecimal number to decimal number.

Solution: In this example fraction number and non-fraction number both are combined together, so you can convert both numbers separately and the converted numbers are combined together to get the final result.

Non-fraction number = 2A6C, this number can be converted as follows:

$2 \times 16^3 + A \times 16^2 + 6 \times 16^1 + C \times 16^0$
$2 \times 4096 + 10 \times 256 + 6 \times 16 + 12 \times 1$ (where A = 10 and C = 12)
$8192 + 2560 + 96 + 12$
10860

Fraction number = 0.2AD, this number can be converted as follows:

$2 \times 16^{-1} + A \times 16^{-2} + D \times 16^{-3}$
$2 \times 1/16 + 10 \times 1/16^2 + 13 \times 1/16^2$ (where A = 10 and D = 13)
$2 \times 1/16 + 10 \times 1/256 + 13 \times 1/4096$
$2 \times 0.0625 + 10 \times 0.003906 + 13 \times 0.000244$
$0.125 + 0.03906 + 0.03172$
0.54732

Now, both can be combined together $(10860.54732)_{10}$, so this is the decimal number equivalent to hexadecimal number $(2A6C.2AD)_{16}$.

The result is $(10860.54732)_{10}$.

2.16 HEXADECIMAL TO BINARY CONVERSION

The highest element of hexadecimal number is fifteen (F (15)). The minimum four bits required to represent the 15(F) in binary. Each digit of hexadecimal number is represented in four bits in binary. Each digit of hexadecimal number represented binary number; all together is the hexadecimal conversion to binary number.

Number Systems **29**

EXAMPLE 2.36 Convert hexadecimal number $(234)_{16}$ to binary number.

Solution: The individual digit of hexadecimal number 234 is 2, 3 and 4. Represent each digit in four-bit representation as follows:

The four-bit binary representation of 2 is

Weight	W4 = 8	W3 = 4	W2 = 2	W1 = 1
Binary	B1 = 0	B2 = 0	B3 = 1	B4 = 0

Total weight = W4 × B1 + W3 × B2 + W2 × B3 + W1 × B4

Total weight = $8 \times 0 + 4 \times 0 + 2 \times 1 + 1 \times 0 = 0 + 0 + 2 + 0 = 2$, so binary representation of 2 is 0010, similarly

Binary representation of 3 is $+8 \times 0 + 4 \times 0 + 2 \times 1 + 1 \times 1 = 3 =$ **0011**

Binary representation of 4 is $8 \times 0 + 4 \times 1 + 2 \times 0 + 1 \times 0 = 4 =$ **0100**

The binary representation of octal number $(234)_{16}$ is **001000110100**

OR

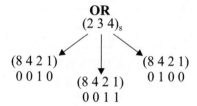

Result is **0010 0011 0100**.

2.16.1 Hexadecimal Fraction to Binary Conversion

The conversion of hexadecimal fraction number to binary fraction number is done by representing each digit of hexadecimal fraction number into the four-bit representation and all digit representation combined together and put decimal point as prefix to get the result.

EXAMPLE 2.37 Convert hexadecimal number $(0.234)_{16}$ to binary number.

Solution: Four bits binary representation of 2 is

Weight	W4 = 8	W3 = 4	W2 = 2	W1 = 1
Binary	B1 = 0	B2 = 0	B3 = 1	B4 = 0

Total weight = W4 × B1 + W3 × B2 + W2 × B3 + W1 × B4

Total weight = $8 \times 0 + 4 \times 0 + 2 \times 1 + 1 \times 0 = 0 + 0 + 2 + 0 = 2$, so binary representation of 2 is 0010, similarly

Binary representation of 3 is $8 \times 0 + 4 \times 0 + 2 \times 1 + 1 \times 1 = 3 =$ **0011**

Binary representation of 4 is $8 \times 0 + 4 \times 1 + 2 \times 0 + 1 \times 0 = 4 =$ **0100**

The binary representation of octal number $(0.234)_{16}$ is **0.001000110100**

30 Digital Logic Design

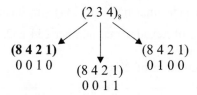

Result is **0.0010 0011 0100**.

EXAMPLE 2.38 Convert hexadecimal number $(234.23)_{16}$ to binary number.

Solution: In this example, the hexadecimal number is consisting of hexadecimal fraction and hexadecimal non-fraction.

Hexadecimal non-fraction = 234, the conversion of $(234)_{16}$ hexadecimal number to binary is as follows:

Three-bit binary representation of 2 is

Weight	W4 = 8	W3 = 4	W2 = 2	W1 = 1
Binary	B1 = 0	B2 = 0	B3 = 1	B4 = 0

Total weight = W4 × B1 + W3 × B2 + W2 × B3 + W1 × B4

Total weight = 8 × 0 + 4 × 0 + 2 × 1 + 1 × 0 = 0 + 0 + 2 + 0 = 2, so binary representation of 2 is 0010, similarly

Binary representation of 3 is 8 × 0 + 4 × 0 + 2 × 1 + 1 × 1 = 3 = 0011
Binary representation of 4 is 8 × 0 + 4 × 1 + 2 × 0 + 1 × 0 = 4 = 0100
The binary representation of octal number $(234)_{16}$ is 001000110100
Similarly,

Hexadecimal fraction number = 0.23, the conversion of $(0.23)_{16}$ hexadecimal number to binary is as follows:

Four-bit binary representation of 2 is

Weight	W4 = 8	W3 = 4	W2 = 2	W1 = 1
Binary	B1 = 0	B2 = 0	B3 = 1	B4 = 0

Total weight = W4 × B1 + W3 × B2 + W2 × B3 + W1 × B4

Total weight = 8 × 0 + 4 × 0 + 2 × 1 + 1 × 0 = 0 + 0 + 2 + 0 = 2, so binary representation of 2 is **0010**, similarly

Binary representation of 3 is 8 × 0 + 4 × 0 + 2 × 1 + 1 × 1 = 3 = 0011

The combined binary representation of digits 2 and 3 is 0010 0011, put decimal point before combined number, and then the result is 0.00100011.

Now, combined non-hexadecimal fraction conversion and fraction hexadecimal conversion together is result for conversion of $(234.23)_{16}$ to binary number. Result is **001000110100.00100011**.

2.17 BINARY TO HEXADECIMAL CONVERSION

Binary to hexadecimal conversion is done by making group of four bits from left to right and convert each group to decimal number and combine all together to get the hexadecimal equivalent to the given binary number. Here we are making group of four bits only because minimum bits required for representing the highest element (F(16)) in hexadecimal number. If you are making a group from left to right, the final group having only two or one digit, put two zeros or three zeros to make four digit before the digit in a group. Putting zero before digits in the group will not affect the number.

EXAMPLE 2.39 Convert binary number $(11011011)_2$ to hexadecimal number.

Solution: The four digits from left to right be 1011, Group-1.

The four digits from Group-1 is 1101 Group-2, now convert the each group in decimal number by adding the weight of each digit in a group, Group-1 = 1011, the weight from right to left is 1, 2, 4, 8; multiply first digit by 1, second digit by 2, third digit by 4 and fourth digit by 8 and the sum of these all are the value of Group-1, similarly, find the value of Group-2, the result is Group-2Group1.

The value of Group-1 is $8 \times 1 + 4 \times 0 + 2 \times 1 + 1 \times 1 = 8 + 0 + 2 + 1 = 11 = B$
The value of Group-2 is $8 \times 1 + 4 \times 1 + 2 \times 0 + 1 \times 1 = 8 + 4 + 0 + 1 = 13 = D$
The result is Group-2Group-1 = $(DB)_{16}$

Therefore $(DB)_{16}$ is hexadecimal number equivalent to binary number $(11011011)_2$.
Result is $(DB)_{16}$.

2.17.1 Binary Fraction to Hexadecimal Conversion

The binary fraction to hexadecimal conversion is done by making group of four binary digits from right to left. Convert each group to decimal number and after conversion combine all groups together to get the result. If in the last group number digit is less than four digits, you add a zero to the last digit of the group to make four digits in a group. Putting zero to the last digit of the group will not affect the number.

EXAMPLE 2.40 Convert binary number $(0.11011011)_2$ to hexadecimal number.

Solution: Making group of three digits from right to left as follows:

Group-1 = 1101
Group-2 = 1011

Now, find the decimal value of each group:

The value of Group-1 = 1101 = $8 \times 1 + 4 \times 1 + 2 \times 0 + 1 \times 1 = 8 + 4 + 0 + 1 = 13 = D$
The value of Group-2 = 1011 = $8 \times 1 + 4 \times 0 + 2 \times 1 + 1 \times 1 = 8 + 0 + 2 + 1 = 11 = B$
The value of Group-3 = 110 = $4 \times 1 + 2 \times 1 + 1 \times 0 = 4 + 2 + 0 = 6$

All groups combine together are .Group-1Group-2=.DB, so the hexadecimal number $(.DB)_{16}$ is equivalent to binary number $(0.11011011)_2$.

32 Digital Logic Design

EXAMPLE 2.41 Convert binary number $(11101.1101)_2$ to hexadecimal number.

Solution: In this example fraction and non-fraction binary numbers are combined. Both numbers are converted separately and the result is combined together.

Non-fraction binary number = 11101

Make the group of four digits from left to right. Group-1 = 1101, Group-2 = 1, in Group-2 only one digit is involved so put three zeros before the digit to make a group of four like Group-2 = 0001.

The value of Group-1 = 1101 = $8 \times 1 + 4 \times 1 + 2 \times 0 + 1 \times 1 = 8 + 4 + 0 + 1 = 13 = D$
The value of Group-2 = 0001 = $8 \times 0 + 4 \times 0 + 2 \times 0 + 1 \times 1 = 0 + 0 + 0 + 1 = 1$

All groups together are Group-2Group-1 = 1D, it is hexadecimal number equivalent to binary number 11101.

Fraction binary number = 0.1101

In this number make group of four digits from right to left. Group-1 = 1101.
The value of Group-1 = 1101 = $8 \times 1 + 4 \times 1 + 2 \times 0 + 1 \times 1 = 8 + 4 + 0 + 1 = 13 = .D$. Now, put both fraction number and non-fraction number together to get the result. The both combined number is $(1D.D)_{16}$, this hexadecimal number is equivalent to binary number $(11101.1101)_2$.

2.18 DECIMAL TO HEXAL NUMBER CONVERSION

The conversion of decimal to hexal number means changing the base 10 of decimal to base 6 of hexal number. Divide the decimal number by 6 which is the base of hexal number and record the remainder, do it until the number is zero. Write the remainder from bottom to top, you will get the result.

EXAMPLE 2.42 Convert the decimal number $(84)_{10}$ to hexal number.
Solution:

Divisor	Number	Remainder
6	84	Initial
6	14	0 (LSB)
6	2	2
6	0	2 (MSB)

The remainder in the above table is 022; the reverse of remainder is 220, which is the hexal equivalent of decimal number 84.

2.18.1 Decimal Fraction to Hexal Conversion

Multiply the fraction decimal number by six (6), and record the number before the decimal point, do this until the number is zero or number starts repeating or up to given iteration in the question or you can assume eight or nine iterations if the number neither repeating nor zero. The recorded number is the result.

EXAMPLE 2.43 Convert the decimal number 0.98 to hexal number.

Solution:

Number	Result
$0.98 \times 6 = 5.88$	5
$0.88 \times 6 = 5.28$	5
$0.28 \times 6 = 1.68$	1
$0.68 \times 6 = 4.08$	4
$0.08 \times 6 = 0.48$	0
$0.48 \times 6 = 2.88$	2
$0.88 \times 6 = 5.28$	Repeat

In this type of example the number is repeating after certain iteration, so you can stop. The result of this example is $(0.551402)_6$.

EXAMPLE 2.44 Convert the decimal number 0.85 to hexal number.

Solution:

Number	Result
$0.85 \times 6 = 5.1$	5
$0.1 \times 6 = 0.6$	0
$0.6 \times 6 = 3.6$	3
$0.6 \times 6 = 3.6$	Repeat

In this example the number is repeating from 0.6 so you can stop and write the result from top to bottom without including the number repeating at 0.6. The result is $(0.503)_6$.

EXAMPLE 2.45 Convert the decimal number 0.25 to hexal number.

Solution:

Number	Result
$0.25 \times 6 = 1.5$	1
$0.5 \times 6 = 3.0$	3
$0.0 \times 6 = 0.0$	Stop

The result is $(0.13)_6$.

EXAMPLE 2.46 Convert the decimal number 0.87 to hexal number.

Solution:

Number	Result
$0.87 \times 6 = 5.22$	5
$0.22 \times 6 = 1.32$	1
$0.32 \times 6 = 1.92$	1
$0.92 \times 6 = 5.52$	5
$0.52 \times 6 = 3.12$	3
$0.12 \times 6 = 0.72$	0
$0.72 \times 6 = 4.32$	4
$0.32 \times 6 = 1.92$	Repeat

The result is $(0.5115304)_6$.

EXAMPLE 2.47 Convert the decimal number 85.85 to hexal number.

Solution: In this example the fraction part and non-fraction part are combined. You can separately convert the fraction part and non-fraction part. After separately conversion of both parts you combine both together to get the result.

Non-fraction part: 85, the conversion of this number is as follows:

Divisor	Number	Remainder
6	85	Initial
6	14	1 (LSB)
6	2	2
6	0	2 (MSB)

The remainder in the above table is 122; the reverse of remainder is 221, which is the hexal equivalent of decimal number 85.

Fraction part: 0.85, the conversion of this number is as follows:

Number	Result
$0.85 \times 6 = 5.1$	5
$0.1 \times 6 = 0.6$	0
$0.6 \times 6 = 3.6$	3
$0.6 \times 6 = 3.6$	Repeat

In this example the number is repeating from 0.6 so you can stop and write the result from top to bottom without including the number repeating at 0.6. The result is 0.503. Now, you can combine both parts together as $(221.503)_6$, this is hexal number equivalent of decimal number $(85.85)_{10}$.

2.19 HEXAL NUMBER TO DECIMAL NUMBER CONVERSION

Hexal number to decimal conversion means changing the base 6 of hexal number to the base 10 of decimal number. For the conversion from hexal to decimal, first count the total number of digits evolves in hexal number. Multiply first digit of hexal number by 6 to the power of total number of digits minus 1 and add it to the next digit then multiply by 6 to the power of total number of digits minus 2, and do it until the last digit of hexal number, the summation of all digits is the result. Let X is a hexal number consisting of $a_0, a_1, a_2, a_3, a_4, \cdots, a_n$, and Y is decimal equivalent of hexal number X. The general formula for converting X into Y is as follows:

$Y = a_0 \times 6^n + a_1 \times 6^{n-1} + a_2 \times 6^{n-2} + a_3 \times 6^{n-3} + a_4 \times 6^{n-4} + \cdots + a_n \times 6^{n-n}$, this can also be written in short for as $Y = \sum_{i=0}^{n} a_i \times 6^{n-i}$.

EXAMPLE 2.48 Convert the hexal number $(254)_6$ to decimal number.

Solution: $2 \times 6^2 + 5 \times 6^1 + 4 \times 6^0$
$2 \times 36 + 5 \times 6 + 4 \times 1$
$72 + 30 + 4$
106

Result is $(106)_{10}$.

2.19.1 Hexal Fraction to Decimal Conversion

Multiply each digit of hexal fraction number by 6 to the power minus 1 and add it to next digit then multiply by 6 to the power minus 2, continuously do it until the end of fraction digit. Let X is the hexal fraction number and its digits are $a_0, a_1, a_2, a_3, a_4, \cdots, a_n$, and Y is decimal number equivalent to hexal fraction number X. The general formula for hexal fraction to decimal conversion is $a_1 \times 6^{-1} + a_2 \times 6^{-2} + a_3 \times 6^{-3} + a_4 \times 6^{-4} + a_5 \times 6^{-5} + \cdots + a_n \times 6^{-n}$, where $n = 1, 2, \cdots, n$. It can be also represented in short form as decimal number $Y = \sum_{i=1}^{n} a_i \times 6^{-n}$.

EXAMPLE 2.49 Convert $(0.253)_6$ hexal number to decimal number.

Solution: $2 \times 6^{-1} + 5 \times 6^{-2} + 3 \times 6^{-3}$
$2 \times 1/6 + 5 \times 1/62 + 3 \times 1/63$
$2 \times 1/6 + 5 \times 1/36 + 3 \times 1/216$
$2 \times 0.1667 + 5 \times 0.2778 + 3 \times 0.0046$
$0.3334 + 1.389 + 0.0138$
1.7362

The result is $(1.7362)_{10}$.

EXAMPLE 2.50 Convert $(254.233)_6$ hexal number to decimal number.

Solution: $2 \times 6^2 + 5 \times 6^1 + 4 \times 6^0 + 2 \times 6^{-1} + 3 \times 6^{-2} + 3 \times 6^{-3}$
$2 \times 36 + 5 \times 6 + 4 \times 1 + 2 \times 1/6 + 3 \times 1/6^2 + 3 \times 1/6^3$

36 Digital Logic Design

$72 + 30 + 4 + 2 \times 1/6 + 3 \times 1/36 + 3 \times 1/216$
$106 + 2 \times 0.1667 + 3 \times 0.2778 + 3 \times 0.0046$
$106 + 0.3334 + 0.8334 + 0.0138$
107.1806

The result is $(107.1806)_{10}$.

2.20 DECIMAL NUMBER TO NONAL NUMBER CONVERSION

The conversion of decimal to nonal number means changing the base 10 of decimal to base 9 of nonal number. Divide the decimal number by 9 which is the base of nonal number and record the remainder, do it until the number is zero. Write the remainder from bottom to top, you will get the result.

EXAMPLE 2.51 Convert the decimal number $(84)_{10}$ to nonal number.

Solution:

Divisor	Number	Remainder
9	84	Initial
9	9	3 (LSB)
9	1	0
9	0	1 (MSB)

The remainder in the above table is 301; the reverse of remainder is 103, which is the nonal equivalent of decimal number 84.

2.20.1 Decimal Fraction to Nonal Conversion

Multiply the fraction decimal number by nine (9) and record the number before the decimal point, do this until the number is zero or number starts repeating or up to given iteration in the question or you can assume eight or nine iteration if the number neither repeating nor zero. The recorded number is the result.

EXAMPLE 2.52 Convert the decimal number 0.92 to nonal number.

Solution:

Number	Result
$0.92 \times 9 = 8.28$	8
$0.28 \times 9 = 2.52$	2
$0.52 \times 9 = 4.68$	4
$0.68 \times 9 = 6.12$	6
$0.12 \times 9 = 1.08$	1
$0.08 \times 9 = 0.72$	0
$0.72 \times 9 = 6.48$	6

In this type of example the number is neither repeating nor zero, so you can stop after some iteration. The result of this example is $(0.8246106)_9$.

2.21 NONAL NUMBER TO DECIMAL NUMBER CONVERSION

Nonal number to decimal conversion means changing the base 9 of nonal number to the base 10 of decimal number. For the conversion from nonal to decimal, first count the total number of digits evolves in nonal number. Multiply first digit of nonal number by 9 to the power of total number of digit minus 1 and add it to the next digit then multiply by 9 to the power of total minus 2, and do it until the last digit of nonal number, the summation of all digits is the result. Let X is a nonal number consisting of $a_0, a_1, a_2, a_3, a_4, \cdots, a_n$, and Y is a decimal number equivalent of nonal number X. The general formula for converting X into Y is as follows:
$Y = a_0 \times 9^n + a_1 \times 9^{n-1} + a_2 \times 9^{n-2} + a_3 \times 9^{n-3} + a_4 \times 9^{n-4} + \cdots + a_n \times 9^{n-n}$, this can also be written in short for as $Y = \sum_{i=0}^{n} a_1 \times 9^{n-i}$.

EXAMPLE 2.53 Convert the nonal number $(254)_9$ to decimal number.
Solution: $2 \times 9^2 + 5 \times 9^1 + 4 \times 9^0$
$2 \times 81 + 5 \times 9 + 4 \times 1$
$162 + 45 + 4$
211

The result is $(211)_{10}$.

2.21.1 Nonal Fraction to Decimal Conversion

Multiply each digit of nonal fraction number by 9 to the power minus 1 and add it to the next digit then multiply by 9 to the power minus 2, continuously do it until the end of fraction digit. Let X is the nonal fraction number and its digits are $a_0, a_1, a_2, a_3, a_4, \cdots, a_n$, and Y is the decimal number equivalent to nonal fraction number X. The general formula for nonal fraction to decimal conversion is $a_1 \times 9^{-1} + a_2 \times 9^{-2} + a_3 \times 9^{-3} + a_4 \times 9^{-4} + a_5 \times 9^{-5} + \cdots + a_n \times 9^{-n}$, where $n = 1, 2, \cdots n$. It can be also represented in short form as decimal number $Y = \sum_{i=1}^{n} a_i \times 9^{-n}$.

EXAMPLE 2.54 Convert $(0.253)_9$ nonal number to decimal number.
Solution: $2 \times 9^{-1} + 5 \times 9^{-2} + 3 \times 9^{-3}$
$2 \times 1/9 + 5 \times 1/9^2 + 3 \times 1/9^3$
$2 \times 1/9 + 5 \times 1/81 + 3 \times 1/729$
$2 \times 0.1111 + 5 \times 0.0123 + 3.0014$
$0.2222 + 0.0615 + 0.0042$
0.2879

The result is $(0.2879)_{10}$.

EXAMPLE 2.55 Convert $(254.253)_9$ nonal number to decimal number.

Solution: $2 \times 9^2 + 5 \times 9^1 + 4 \times 9^0 + 2 \times 9^{-1} + 5 \times 9^{-2} + 3 \times 9^{-3}$
$2 \times 81 + 5 \times 9 + 4 \times 1 + 2 \times 1/9 + 5 \times 1/9^2 + 3 \times 1/9^3$
$162 + 45 + 4 + 2 \times 1/9 + 5 \times 1/81 + 3 \times 1/729$
$162 + 45 + 4 + 2 \times 0.1111 + 5 \times 0.0123 + 3 \times 0.0014$
$211 + 0.2222 + 0.0615 + 0.0042$
$211 + 0.2879$
211.2879

The result is $(211.2879)_{10}$.

EXERCISES

Short Answer Questions

1. Write first 30 numbers in binary, octal, hexadecimal.
2. Find the value of X, if $(234)_{10} = (X)_7$.
3. Convert the binary number 1101101 to the decimal, octal and hexadecimal number.
4. Convert the octal number $(456)_8$ to hexadecimal and decimal number.
5. Convert the following to the base indicated:
   (a) $(42)_{10}$ to $(\ )_8$
   (b) $(50.25)_8$ to $(\ )_{16}$
   (c) $(11101.1101)_2$ to $(\ )_8$
   (d) $(A2B.5F)_{16}$ to $(\ )_{10}$
6. Convert the number $(456)_{10}$ to base 4, base 6, base 8, base 9 and base 16.
7. Convert $(23)_6$ to base 8, base 9 and base 12.
8. If $(85)_{10} = (221)_x$, find the value of x.
9. Convert following hexadecimal numbers to binary, octal and hexadecimal number.
   (a) $(2A45)_{16}$
   (b) $(45.50D)_{16}$
   (c) $(6BC)_{16}$
   (d) $(6.BC)_{16}$
10. If $(x.y)_{16} = (111101.1100)_2$. Find the value of x and y.
11. $x = 10860$ and $y = 2A6C$. $x = y$ if the base of y is 16 find the base of x.
12. Find the missing terms in the following series: 11, 12, 13, 14, 15, 16, 17,___, ___
13. Find the missing terms in the following series: 10, 12, 13, 20, 21, 22, 23, __, __
14. Find the missing terms in the following series: A0, A1, A2, A3, __, __, A6, __, A8, __.
15. Find the missing terms in the following series:___, ____, 20, 21, 22, 23, 24, 25, 26, 17.

Long Answer Questions

1. Convert the following octal numbers to binary decimal and hexadecimal numbers.
   (a) $(65)_8$
   (b) $(70.34)_8$
   (c) $(45.52)_8$
   (d) $(48.3)_8$

2. Convert the following binary numbers to octal, decimal hexadecimal numbers.
   (a) $(11101101.11011)_2$
   (b) $(10011101)_2$
   (c) $(110110001.111101)_2$
   (d) $(10000111.1101)_2$
3. Which of the following numbers are not octal numbers: 245, 458, 345, 295, 573.
4. Find the decimal number from the following numbers: 234, 593, 234, 567, 203.
5. What are the possible bases of the following numbers: 234, 567, 876.
6. Convert the following decimal numbers to hexal and nonal numbers.
   (a) $(12.34)_{10}$
   (b) $(54.56)_{10}$
   (c) $(564)_{10}$
   (d) $(503)_{10}$
7. Convert the following hexal numbers to decimal numbers.
   (a) $(45)_6$
   (b) $(53.25)_6$
   (c) $(102.34)_6$
   (d) $(4.5)_6$
7. Convert the following nonal numbers to decimal numbers.
   (a) $(55)_9$
   (b) $(43.25)_9$
   (c) $(112.34)_9$
   (d) $(14.5)_9$

Multiple Choice Questions

1. Bits stand for
   (a) Binary Digit
   (b) Binary Information
   (c) Binary Information Technology
   (d) None
2. The radix of a number system x is 16, the x represents
   (a) Octal number
   (b) Hexadecimal number
   (c) Decimal number
   (d) Septa number
3. The octal representation of binary number 1101.110 is
   (a) 15.3
   (b) 61.6
   (c) 14.3
   (d) 61.3
4. If $x = (101)_2$ and $y = 5$ when x converted to y. What is the base of y?
   (a) 4
   (b) 5
   (c) 8
   (d) 16
5. Convert the decimal number 53 to 6-bit binary.
   (a) 110101
   (b) 101011
   (c) 110010
   (d) 111010
6. Convert binary 111110110 to hexadecimal.
   (a) 2E6
   (b) EE3
   (c) BC6
   (d) 1E6
7. Convert hexadecimal number BC to binary.
   (a) 10111100
   (b) 11010011
   (c) 10000111
   (d) 11100110

8. If the hexadecimal number B2B is converted to binary number, then the number of 1 present in the binary number is
   (a) 4 (b) 6
   (c) 7 (d) 8

9. The radix of number X is 6, if $X = 5$. What is octal equivalent of X?
   (a) 5 (b) 6
   (c) 8 (d) 4

10. The missing terms in the following series: 11, 12, 13, 14, 15, 16, 17, 20, ___, is
    (a) 18 (b) 20
    (c) 21 (d) 22

11. The missing terms in the following series: 11, 12, 13, 14, 15, 21, 22, 23, ___, is
    (a) 24 (b) 25
    (c) 34 (d) 36

12. The number 24 converted to 20, if number 20 is in decimal number then the base of number 24 is
    (a) 8 (b) 4
    (c) 16 (d) 10

13. The missing term in the following series 0, 1, 10, 11, ___, 101 is
    (a) 110 (b) 111
    (c) 100 (d) 010

14. The formula for conversion of hexal number to decimal number conversion is
    (a) $\sum_{i=0}^{n} a_1 \times 6^{n-i}$ (b) $\sum_{i=0}^{n} a_1 \times 3^{n-i}$
    (c) $\sum_{i=0}^{n} a_1 \times 7^{n-i}$ (d) $\sum_{i=0}^{n} a_1 \times 8^{n-i}$

15. The base X and Y are 10 and 16, respectively. The general formula for converting X into Y is
    (a) $= \sum_{i=0}^{n} a_1 \times 16^{n-i}$ (b) $= \sum_{i=0}^{n} a_1 \times 16^{n-i}$
    (c) $= \sum_{i=0}^{n} X_1 \times 16^{n-i}$ (d) $= \sum_{i=0}^{n} a_1 \times 6^{n-i}$

Answers

1. (a)	2. (b)	3. (a)	4. (c)	5. (a)	6. (d)
7. (a)	8. (c)	9. (a)	10. (c)	11. (a)	12. (a)
13. (c)	14. (a)	15. (c)			

BIBLIOGRAPHY

Sterling, M.J., *Algebra II Workbook for Dummies*.

CHAPTER 3

Data and Information Representation

3.1 INTRODUCTION

Data and information representation mean using symbol to represent data as well as information. The symbol is nothing but zero and one. The data and information represented in the form of zero and one are known as data and information representation. The computer can understand one as high voltage and zero as low voltage. But human being can understand English like language easily, so the textual as well as numerical data and information are represented in computer understanding format, means zero and one. Generally in a digital system or digital computer, a more precise representation of a signal is obtained by using more binary digits for representation.

3.2 DATA AND INFORMATION

The data and information play an important role in a computer system. The processed data is known as information. The information may be data for the further processing. The data is processed under certain rules and regulation.

3.3 NUMBER NORMALIZATION

A number is said to be normalized if decimal point is followed by nonzero number. A number in which decimal number is followed by zero means the number is not normalized. Let us consider an example number 0.000234, in this number decimal is followed by zero so this number is not normalized. Consider another number .2034 in this number decimal point is not followed by zero so you can say this number is normalized number. The number is generally normalized in the case of floating point arithmetic operation. Normalized number is required when you are going to compare two numbers.

For the efficient use of the bits available for the significant, you can shift the bit to the left until all leading zero (0)'s out (as there is no contribution to the precision). The value is kept unchanged by adjusting the exponent accordingly. The most significant bit is always one (1). There is no need to show it explicitly. The significant is further shifted to the left by one (1) bit to gain one more bit for precision. The first bit one (1) is implicit before the decimal point. The actual value can be represented as follows:

$$(-1)^s \times (1 + m) \times 2^e$$

To avoid possible confusion, the default normalization does not assume this implicit one (1) unless otherwise specified in the following. If zero is represented by each zero's, it is not normalized. The exponent is represented in 2's complements representation for representing negative number.

EXAMPLE 3.1 The binary number $y = 0.000011101101101$ can be represented in 16-bits floating-point form in this way, one bit is reserved for sign bit, 5 bits are reserved for exponent and 10 bits for mantissa.

Solution:

$y = 0.000011101101101 \times 2^0$ | 0 | 00000 | 0000111011 |

$y = 0.00011101101101 \times 2^{-1}$ | 0 | 11111 | 0001110110 |

$y = 0.0011101101101 \times 2^{-2}$ | 0 | 11110 | 0011101101 |

$y = 0.011101101101 \times 2^{-3}$ | 0 | 11101 | 0111011011 |

$y = 0.11101101101 \times 2^{-4}$ | 0 | 11100 | 1110110110 |

$y = 1.1101101101 \times 2^{-5}$ | 0 | 11011 | 1101101101 |

Now the y is normalized, and the implied is 1.0.

EXAMPLE 3.2 Normalized the number 0.0002345.

Solution: The decimal point is followed by non-zero number. Shift the decimal point to right where number is not zero without affecting the value of number. For example, $0.0002345 = 0.2345 \times 10^{-3}$, so the normalized number is 0.2345×10^{-3}. This is the direct solution. This can also be done as follows:

Suppose $x = 0.0002345$
             $x = 0.0002345 \times 10^0$
             $x = 0.002345 \times 10^{-1}$
             $x = 0.02345 \times 10^{-2}$
             $x = 0.2345 \times 10^{-3}$

Now the value of x is the normalized number.

3.4 FLOATING POINT REPRESENTATION

Computers which work with real arithmetic use floating point representation mean representing a number in mantissa and exponent part. Consider a number 23.456, this number can be divided into two part mantissa and exponent by changing the format of number, this number can be written as 0.23456×10^2. In this number, 23456 is mantissa part and 10^2 is exponent part. This number can also be written as 0.23456E2. The following are some cases of decimal number: 45.1234578943..., 3.86732..., 7.5678..., 5.0032×10^{12}, 8.456×10^{21}. When the programmer is writing a program in any language, they can express a floating point number -23.425×10^{-7} as -23.425E-7. This can be defined in general term as $\pm m \times b^e$ where m is mantissa or significant, e is exponent and b indicates the base of number, in the case of decimal $b = 10$, in the case of binary $b = 2$ and in the case octal $b = 8$, so on.

3.4.1 Floating Point Representation of 32 Bits Binary Number

Let consider 32-bit word which is used to represent a floating-point number in Figure 3.1.

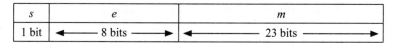

Figure 3.1 32-bit floating point representation.

Here 1 bit is reserved for sign bit, 8 bits are reserved for exponent and 23 bits are reserved for mantissa bits. This can be represented as follows:

$$(-1)^s \times m \times 2^e$$

A floating number is consisting of two parts, one part before decimal number and another part after decimal number, e.g. 12.34. The floating point number is represented in 32-bits floating point representation is as follows: Take 24 bits for mantissa part including 1 bit for sign, and 8-bits for exponent including sign bit. The pictorial representation is given in Figure 3.2.

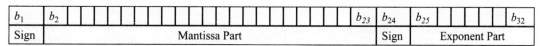

Figure 3.2 Pictoral representation of 32-bit floating point number.

Where $b_1, b_2, b_3, ..., b_{32}$ are bits position. If the number of binary digit is less than 23 in mantissa part put zero(s) from left to right to make 23 binary digits. In the case of binary digit less than seven in exponent part, put zero(s) from right to left to make seven digit exponent parts. Doing in this way the number will not be affected.

EXAMPLE 3.3 Represent -2.35 in 32-bit binary floating point representation.
Solution: The binary equivalent of decimal number 2 is 10. The binary equivalent of 0.35 is

$$0.35 \times 2 = 0.7 \quad 0$$
$$0.7 \times 2 = 1.4 \quad 1$$
$$0.4 \times 2 = 0.8 \quad 0$$
$$0.8 \times 2 = 1.6 \quad 1$$
$$0.6 \times 2 = 1.2 \quad 1$$
$$0.2 \times 2 = 0.4 \quad \text{Repeat}$$

Therefore $0.35 = .01011$, both taken together is 10.01011 this can be represented scientifically as 0.1001011×2^{-2}.

Mantissa part is -0.1001011 exponent part is $-2 = 1, 10$ (in signed magnitude form). This can be represented in 32-bit binary floating point in Figure 3.3.

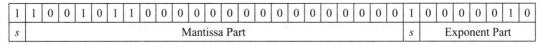

1	1	0	0	1	0	1	1	0	0	0	0	0	0	0	0	0	0	0	0	0	0	0	0	1	0	0	0	0	0	1	0
s	Mantissa Part																									s	Exponent Part					

Figure 3.3 32-bit floating point representation of -2.35.

3.4.2 Floating Point Representation of 16 Bits Binary Number

Let consider 16-bit word is used to represent a floating-point number. The floating-point number is represented in 16-bits floating point representation is as follows: Take 9 bits for mantissa part including 1 bit for sign, and 7 bits for exponent including sign bit. The pictorial representation is given in Figure 3.4.

b_1	b_2	b_3	b_4	b_5	b_6	b_7	b_8	b_9	b_{10}	b_{11}	b_{12}	b_{13}	b_{14}	b_{15}	b_{16}
Sign	Mantissa Part								Sign	Exponent Part					

Figure 3.4 16-bit floating point representation.

Where $b_1, b_2, b_3, \ldots, b_{16}$ are bits position. If the number of binary digit is less than 8 in mantissa part put zero from left to right to make 8 binary digits. In the case of binary digit less than 6 in exponent part, put zero from right to left to make 6-digit exponent parts. Doing in this way the number will not be affected.

EXAMPLE 3.4 Represent -2.35 in 16-bit binary floating-point representation.

Solution: The binary equivalent of decimal number 2 is 10. The binary equivalent of 0.35 is

$$0.35 \times 2 = 0.7 \quad 0$$
$$0.7 \times 2 = 1.4 \quad 1$$
$$0.4 \times 2 = 0.8 \quad 0$$
$$0.8 \times 2 = 1.6 \quad 1$$
$$0.6 \times 2 = 1.2 \quad 1$$
$$0.2 \times 2 = 0.4 \quad \text{Repeat}$$

Therefore 0.35 = 0.01011, both taken together is 10.01011 this can be represented scientifically as 0.1001011×2^{-2}.

Mantissa part is −0.1001011 exponent part is −2 = 1, 10 (in signed magnitude form). This can be represented in 32-bit binary floating point in Figure 3.5.

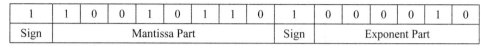

Figure 3.5 16-bit floating point representation of −2.35.

3.4.3 Range of 16 Bits Binary Number

The range of 16-bit binary number depends on the number bit allotted to mantissa part and exponent part. Assume 9 bits out of 16-bits binary number for mantissa part including sign bit and 7 bits for exponent part including sign bits. The maximum number which can be stored in 16-bit binary number is shown in Figure 3.6.

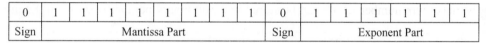

Figure 3.6 Maximum number stored in 16-bit register.

The maximum values that can be stored in 16-bit binary number are $0.111111111e0111111$, where e represents the exponent. This number can be converted into decimal number is as follows:

$0.111111111e0111111$

$= (1 \times 2^{-1} + 1 \times 2^{-2} + 1 \times 2^{-3} + 1 \times 2^{-4} + 1 \times 2^{-5} + 1 \times 2^{-6} + 1 \times 2^{-7} + 1 \times 2^{-8}) \times 2^{0111111}$

$= (2^{-1} + 2^{-2} + 2^{-3} + 2^{-4} + 2^{-5} + 2^{-6} + 2^{-7} + 2^{-8}) \times 2^{(1 \times 26 + 1 \times 25 + 1 \times 24 + 1 \times 23 + 1 \times 22 + 1 \times 21 + 1 \times 20)}$

$= (2^{-1} + 2^{-2} + 2^{-3} + 2^{-4} + 2^{-5} + 2^{-6} + 2^{-7} + 2^{-8}) \times 2^{(26+25+24+23+22+21+20)}$

$= (1 - 2^{-8}) \times 2^{(26-1)}$

$\approx 1 \times 2^{(64-1)}$

$\approx 2^{63}$

$2^{63} = 10^x$

The minimum number which can be stored in 16-bit binary number is shown in Figure 3.7.

0	1	0	0	0	0	0	0	0	1	1	1	1	1	1	1
Sign		Mantissa Part							Sign		Exponent Part				

Figure 3.7 Minimum number stored in 16-bit register.

The minimum values that can be stored in 16-bit binary number are $0.100000000e1111111$, where e represents the exponent. This number can be converted into decimal number is as follows:

46 Digital Logic Design

$0.10000000e1111111$

$= (1 \times 2^{-1} + 0 \times 2^{-2} + 0 \times 2^{-3} + 0 \times 2^{-4} + 0 \times 2^{-5} + 0 \times 2^{-6} + 0 \times 2^{-7} + 0 \times 2^{-8}) \times 2^{1111111}$

$= (2^{-1} + 0 + 0 + 0 + 0 + 0 + 0 + 0) \times 2^{-(1\times 26 + 1\times 25 + 1\times 24 + 1\times 23 + 1\times 22 + 1\times 21 + 1\times 20)}$

$= (2^{-1}) \times 2^{-(26+25+24+23+22+21+20)}$

$= 2^{-1} \times 2^{-(2^6 - 1)}$

$\qquad\qquad = 2^{-1} \times 2^{-63}$

$\qquad\qquad = 2^{-64}$

$2^{-64} = 10^x$

Therefore, the range of 16-bit binary number is 2^{-64} to 2^{63}.

3.4.4 Range of 32 Bits Binary Number

The range of 32-bit binary number depends on the number bit allotted to mantissa part and exponent part. Assume 23 bits out of 32-bits binary number for mantissa part including sign bit and 9 bits for exponent part including sign bits.

The maximum number which can be stored in 32-bit binary number is shown in Figure 3.8.

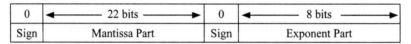

Figure 3.8 Maximum number stored in 32-bit register.

The maximum values that can be stored in 16-bit binary number are $0.11111111111111111111111e011111111$, where e represents the exponent. This number can be converted into decimal number is as follows:

$0.11111111111111111111111e011111111$

$= (1 \times 2^{-1} + 1 \times 2^{-2} + 1 \times 2^{-3} + 1 \times 2^{-4} + 1 \times 2^{-5} + \cdots$
$\qquad + 1 \times 2^{-20} + 1 \times 2^{-21} + 1 \times 2^{-22}) \times 2^{011111111}$

$= (2^{-1} + 2^{-2} + 2^{-3} + 2^{-4} + 2^{-5} + 2^{-6} + 2^{-7} + \cdots + 2^{-22}) \times 2^{(1\times 28 + 1\times 27 + \cdots + 1\times 24 + 1\times 23 + 1\times 22 + 1\times 21 + 1\times 20)}$

$= (2^{-1} + 2^{-2} + 2^{-3} + 2^{-4} + 2^{-5} + 2^{-6} + 2^{-7} + \cdots + 2^{-22}) \times 2^{(28+27+\cdots+24+23+22+21+20)}$

$= (1 - 2^{-22}) \times 2^{(2^8 - 1)}$

$\approx 1 \times 2^{(256-1)}$

$= 2^{255}$

$2^{255} = 10^x$

The minimum number which can be stored in 32-bit binary number is shown in Figure 3.9.

0	1	Remaining bits are zero up to 21 bits	1	All 9 bits are one
Sign		Mantissa Part	Sign	Exponent Part

Figure 3.9 Minimum number stored in 32-bit register.

The minimum values that can be stored in 16-bit binary number are $0.100000000e1111111$, where e represents the exponent. This number which can be converted into decimal number is as follows:

$0.1000000000000000000000e111111111$
$= (1 \times 2^{-1} + 0 \times 2^{-2} + 1 \times 2^{-3} + \cdots + 0 \times 2^{-20} + 0 \times 2^{-21} + 0 \times 2^{-22}) \times 2^{1111111}$
$= (2^{-1} + 0 + 0 + 0 + \cdots + 0 + 0 + 0 + 0) \times 2^{-(1 \times 28 + 1 \times 27 + \cdots + 1 \times 24 + 1 \times 23 + 1 \times 22 + 1 \times 21 + 1 \times 20)}$
$= (2^{-1}) \times 2^{-(28+27+\cdots+24+23+22+21+20)}$
$= (2^{-1}) \times 2^{-(28-1)}$
$= 2^{-1} \times 2^{-255}$
$= 2^{-256}$
$= 2^{-256} = 10^x$

Therefore, the range of 16-bit binary number is 2^{-256} to 2^{255}.

3.5 REPRESENTATION OF NEGATIVE NUMBER

The computer cannot understand negative number. The complement of number and signed magnitude are used to represent a negative number.

3.5.1 Representation of Negative Number in Binary System

There are three methods to represent the negative number in binary number system as given below:

1. Signed Magnitude
2. One's Complement
3. Two's Complement

Signed magnitude

The signed magnitude method is used to represent the negative number in binary number system. In this method the lowest digit that is zero put before the binary number to represent the positive number. The highest digit of binary number is one which is put before the binary number to represent the negative number.

EXAMPLE 3.5 Represent the binary number 101011 as negative number using signed magnitude method.

Solution: The highest digit of binary number system is one. So you can put one before the given number to represent above given number in negative form. The result is 1,101011.

One's complement method

The one's complement is the second method to represent the negative number in binary number system. The highest digit of binary number system is one (1). For finding the one's complement, subtract each digit of given number from one (1) to get the one's complement of a given number. In another way change the digit zero to one and one to zero in given number to get the one's complement.

EXAMPLE 3.6 Represent the binary number 101011 as negative number using one's complement method.

Solution: There are six digits in the given example. Make a number of six digits, all six digits are one. Now subtract the given binary number from you make the six-digit number.

Let x is a number you make it and y is a number which is to be represented in one's complement.

$x = 111111$, $y = 101011$

$$\text{Result} = x - y = 111111 - 101011 = 010100.$$

The number 010100 is one's complement of 101011. The number 010100 is negative representation of number 101011.

Two's complement method

The two's complement is the third method to represent the negative number in binary number system. First find the one's complement of a given number and add one to the obtained one's complement to get the two's complement.

EXAMPLE 3.7 Represent the binary number 101011 as negative number using two's complement method.

Solution: The one's complement of number 101011 is
$x = 111111$, $y = 101011$

$$\text{Result} = x - y = 111111 - 101011 = 010100.$$

Add one to the result obtained above as follows:

$010100 + 000001 = 010101$

The two' complement of the given number is 010101.
The number 010101 is the negative representation in two's complement of a number 101011.

3.5.2 Representation of Negative Number in Octal System

The three methods are used to represent the negative number in octal number system as given below:

1. Signed Magnitude
2. Seven's Complement
3. Eight's Complement

Signed magnitude

In this method the lowest digit that is zero put before the octal number to represent the positive number. The highest digit of octal number is seven which is put before the octal number to represent the negative number.

EXAMPLE 3.8 Represent the octal number 567 as negative number using signed magnitude method.

Solution: The highest digit of octal number system is seven. So you can put seven before the given number to represent above given number in negative form. The result is 7,567.

Seven's complement method

The seven's complement is the second method to represent the negative number in octal number system. The highest digit of octal number system is seven (7). For finding the seven's complement, subtract each digit of given number from seven (7) to get the seven's complement of a given number.

EXAMPLE 3.9 Represent the octal number 567 as negative number using seven's complement method.

Solution: There are three digits in the given example. Make a number of three digits, all three digits are seven. Now subtract the given octal number from you make the three-digit number.

Let x is a number you make it and y is a number which is to be represented in seven's complement.

$x = 777, y = 567$

$$\text{Result} = x - y = 777 - 567 = 210.$$

The number 210 is seven's complement of 567. The number 210 is negative representation of number 567.

Eight's complement method

The eight's complement is the third method to represent the negative number in octal number system. First find the seven's complement of a given number and add one to the obtained seven's complement to get the eight's complement.

EXAMPLE 3.10 Represent the octal number 567 as negative number using eight's complement method.

Solution: The seven's complement of number 567 is
$x = 777, y = 567$

$$\text{Result} = x - y = 777 - 567 = 210.$$

Add one to the result obtained above as follows:

$$210 + 001 = 211$$

The eight's complement of the given number is 211.
The number 211 is the negative representation in eight's complement of a number 567.

3.5.3 Representation of Negative Number in Decimal System

The three methods are also used to represent the negative number in decimal number system as given:

1. Signed Magnitude
2. Nine (9)'s Complement
3. Ten (10)'s Complement

Signed magnitude

In this method the lowest digit that is zero put before the octal number to represent the positive number. The highest digit of decimal number is nine (9) which is put before the decimal number to represent the negative number.

EXAMPLE 3.11 Represent the decimal number 567 as negative number using signed magnitude method.

Solution: The highest digit of decimal number system is nine (9). So you can put nine before the given number to represent above given number in negative form. The result is 9,567.

Nine (9)'s complement method

The nine's complement is the second method to represent the negative number in decimal number system. The highest digit of decimal number system is nine (9). For finding the nine (9)'s complement, subtract each digit of given number from nine (9) to get the nine's complement of a given number.

EXAMPLE 3.12 Represent the decimal number 567 as negative number using nine's complement method.

Solution: There are three digits in the given example. Make a number of three digits, all three digits are nine. Now subtract the given decimal number from you make the three-digit number.

Let x is a number you make it and y is a number which is to be represented in nine's complement.

$x = 999$, $y = 567$

$$\text{Result} = x - y = 999 - 567 = 432.$$

The number 432 is nine's complement of 567. The number 432 is negative representation of number 567.

Ten's (10)'s complement method

The ten's complement is the third method to represent the negative number in decimal number system. First find the nine's complement of a given number and add one to the obtained nine's complement to get the ten's complement.

EXAMPLE 3.13 Represent the decimal number 567 as negative number using ten's complement method.

Solution: The nine's complement of number 567 is
$x = 999$, $y = 567$

$$\text{Result} = x - y = 999 - 567 = 432.$$

Add one to the result obtained above as follows:
$$432 + 001 = 433$$
The ten's complement of the given number is 433.
The number 433 is the negative representation in ten's complement of a number 567.

3.5.4 Representation of Negative Number in Hexadecimal System

The three methods are also used to represent the negative number in hexadecimal number system as given below:

1. Signed Magnitude
2. Fifteen (F)'s Complement
3. Sixteen's Complement

Signed magnitude

In this method the lowest digit that is zero put before the hexadecimal number to represent the positive number. The highest digit of hexadecimal number is fifteen (F) which is put before the decimal number to represent the negative number.

EXAMPLE 3.14 Represent the hexadecimal number 567 as negative number using signed magnitude method.

Solution: The highest digit of decimal number system is fifteen (F). So you can put F before the given number to represent above given number in negative form.
The result is F, 567.

Fifteen (F)'s complement method

The fifteen (F)'s complement is the second method to represent the negative number in hexadecimal number system. The highest digit of hexadecimal number system is fifteen (F). For finding the fifteen (F)'s complement, subtract each digit of given number from fifteen (F) to get the fifteen's complement of a given number.

EXAMPLE 3.15 Represent the hexadecimal number 567 as negative number using fifteen's complement method.

Solution: There are three digits in the given example. Make a number of three digits, all three digits are fifteen (F). Now subtract the given hexadecimal number from you make the three-digit number.
Let x is a number you make it and y is a number which is to be represented in fifteen's complement.
x = FFF, y = 567
Result = $x - y$ = FFF − 567 = A98.

The number A98 is fifteen's complement of 567. The number A98 is negative representation of number 567.

52 Digital Logic Design

Sixteen's complement method

The sixteen's complement is the third method to represent the negative number in hexadecimal number system. First find the fifteen's complement of a given number and add one to the obtained fifteen's complement to get the sixteen's complement.

EXAMPLE 3.16 Represent the hexadecimal number 567 as negative number using sixteen's complement method.

Solution: The fifteen's complement of number 567 is
$x = \text{FFF}, y = 567$

$$\text{Result} = x - y = \text{FFF} - 567 = \text{A98}.$$

Add one to the result obtained above as follows:

$$\text{A98} + 001 = \text{A99}$$

The sixteen's complement of the given number is A99. The number A99 is the negative representation in sixteen's complement of a hexadecimal number 567.

3.6 CODES AND ITS CONVERSION

There are various types of codes which are given as follows:

- BCD code
- Excess-3
- 2421 code
- ASCII
- Gray code
- 8421 code
- $\overline{8421}$ code
- Self-complement code

3.6.1 BCD Code

The BCD code stands for binary coded decimal. This is four-bit code, from zero to nine. For example, 0000, 0001, 0010, 0011, 0100, 0101, 0110, 0111, 1000, 1001.

EXAMPLE 3.17 Convert the number 245 into BCD code.

Solution: Represent each digit of the given number in four-bit representation to get the BCD code of a given number.

The three digits in the given number are 2, 4 and 5. All these digits represented in four bits are as follows:
2 = 0010, 4 = 0100 and 5 = 0101, so BCD code of 245 is 0010 0100 010.

3.6.2 Gray Code

The Gray code is an encoding method of numbers. In Gray code method the adjacent numbers have a single digit differ by one. The Gray code is also known as a "reflected" code, in other words it is also termed as more specifically still, the binary reflected Gray code.

Binary to Gray code conversion

Let x is binary and y is Gray code of x. $x = b_0b_1b_2b_3$. The Gray code equivalent of x is obtained by taking the first bit as it is and next bit b_1 is added to b_0, if carry is generated ignore it, the result is second bit of binary number, add b_1 to b_2 and ignore the carry, the result is third bit of binary number and add b_2 to b_3 by ignoring carry, the result is fourth bit of binary number.

$$x = b_0b_1b_2b_3$$
$$y = b_0(b_0 + b_1)(b_1 + b_2)(b_2 + b_3)$$

EXAMPLE 3.18 Convert the binary number 1011 to Gray code.

Solution: Let x is binary and y is Gray code equivalent of x.

$x = 1011$, here $b_0 = 1$, $b_1 = 0$, $b_2 = 1$, $b_3 = 1$

$y = b_0(b_0 + b_1)(b_1 + b_2)(b_2 + b_3)$, putting the value of b_0, b_1, b_2 and b_3 in y to get the Gray code equivalent of binary.

$y = 1(1 + 0)(0 + 1)(1 + 1) = 1110$, so 1110 is a Gray code of binary 1011.

Gray code to binary conversion

Let x is Gray code and y is binary of x. $x = b_0b_1b_2b_3$ and $y = y_0y_1y_2y_3$. The binary equivalent of x is obtained by taking the first bit as it is and next bit of Gray code is obtained by adding the first bit of Gray code and next bit of binary, the result after ignoring the carry is second bit of Gray code, the third bit of Gray code is brained by adding the second bit of Gray code to third bit of binary after ignoring the carry and so on.

$$x = b_0b_1b_2b_3$$
$$y_0 = b_0, y_1 = b_1 + y_0, y_2 = b_2 + y_1, y_3 = y_2 + b_3$$

Now putting the value of y_0, y_1, y_2 and y_3 in the following equation to get the binary

$$y = y_0y_1y_2y_3, y = b_0(b_1 + y_0)(b_2 + y_1)(y_2 + b_3).$$

EXAMPLE 3.19 Convert the Gray code 1011 to binary.

Solution: Let x is Gray code and y is binary equivalent of x.

$x = 1011$, here $b_0 = 1$, $b_1 = 0$, $b_2 = 1$, $b_3 = 1$ and $y = y_0y_1y_2y_3$

$$y_0 = b_0, y_1 = b_1 + y_0, y_2 = b_2 + y_1, y_3 = y_2 + b_3$$
$$y_0 = b_0 = 1$$
$$y_1 = b_1 + y_0 = 0 + 1 = 1$$
$$y_2 = b_2 + y_1 = 1 + 1 = 0$$
$$y_3 = y_2 + b_3 = 0 + 1 = 1$$

Now putting the value of y_0, y_1, y_2 and y_3 in the following equation to get binary equivalent.

$y = y_0y_1y_2y_3 = 1101$, so 1101 is the binary code of Gray code 1011.

3.6.3 Excess-3 Code

The Excess-3 code is obtained by adding 3 to the given number and representing it to binary number.

EXAMPLE 3.20 Convert the decimal from 0 to 9 into Excess-3 and binary codes.

Solution:

Decimal	Excess-3	Binary
0	0 + 3 = 3	0011
1	1 + 3 = 4	0100
2	2 + 3 = 5	0101
3	3 + 3 = 6	0110
4	4 + 3 = 7	0111
5	5 + 3 = 8	1000
6	6 + 3 = 9	1001
7	7 + 3 = 10	1010
8	8 + 3 = 11	1011
9	9 + 3 = 12	1100

3.6.4 8421 Code

8421 code is a weighted code in which each digit 0 through 9 is represented by a four-bit code. The bit positions in each code are assigned value, from left to right, of 8, 4, 2 and 1. For example, consider a binary number of four bits 1011. The value of first bit is 8 because $8 \times 1 = 8$, the value of second bit in this number is zero because $0 \times 4 = 0$, the value of third bit is 2 because $1 \times 2 = 2$ and for the value of fourth and last bit is 1 because $1 \times 1 = 1$. The total weight of these four bits is $8 + 4 + 2 + 1 = 15$. Therefore, this code represents number 0 to 15.

EXAMPLE 3.21 Convert the decimal from 0 to 9 into 8421 code.

Solution:

Decimal	8421
0	0000
1	0001
2	0010
3	0011
4	0100
5	0101
6	0110
7	0111
8	1000
9	1001

EXAMPLE 3.22 Convert the number 28 into 8421 code.

Solution: There are two digits in the given number, so you can represent each digit in 8421 code and finally combine both codes together to get the answer.

2 is the first digit of number 28, so 2 is represented in 8421 code as 0010.
8 is the second digit of number 28, so 8 is represented in 8421 code as 1000,

The result is 0010 1000.

3.6.5   2421 Code

2421 is a weighted code, the weights of this code are 2, 4, 2 and 1. The decimal number is represented in four bits. The total weights of these four bits are $2 + 4 + 2 + 1 = 9$. Therefore, 2421 code represents decimal number 0 to 9. For example, 0 to 9 decimal numbers represented in 2421 code are given in Table 3.1.

Table 3.1   Decimal number represented in 2421 code

Decimal	2421
0	0000
1	0001
2	0010
3	0011
4	0100
5	0101
6	0110
7	0111
8	1110
9	1111

EXAMPLE 3.23   Convert the number 28 into 2421 code.

Solution: There are two digits in the given number, so you can represent each digit in 2421 code and finally combine both codes together to get the result.

2 is the first digit of number 28, so 2 is represented in 2421 code as 0010.
8 is the second digit of number 28, so 8 is represented in 2421 code as 1110.

The result is 0010 1110.

3.6.6 $8\overline{4}21$ Code

$8\overline{4}21$ code is a weighted code, the weights of this code are 8, –4, 2 and 1. The decimal number is represented in four bits. The total weights of these four bits are $8 - 4 + 2 + 1 = 7$. Therefore, $8\overline{4}21$ code represents decimal number 0 to 7. For example, 0 to 7 decimal numbers represented in $8\overline{4}21$ code are given in Table 3.2.

Table 3.2 Decimal number represented in $8\bar{4}21$ code

Decimal	$8\bar{4}21$
0	0000
1	0001
2	0010
3	0011
4	1100
5	1101
6	1110
7	1111

3.6.7 ASCII Code

ASCII code stands as American Standard Code for Information Interchange. ASCII code is used to representing characters and symbols in electronic form. This code is very useful for communication in large group of peoples. The internet requires even more focus on standards, as it integrates users with different languages. This is seven bits code used for exchange the information. Now 8-bit ASCII code is also introduced. The capital letter of English alphabet A value in ASCII is 65. The letter A is represented in ASCII as seven bits 1000001.

3.6.8 Self-complement Code

A code is said to be self-complement code, if the one's complement of the code is equal to the nine's complement of the same code.

3.7 ERROR DETECTION AND CORRECTION CODE

The message is transmitted from source to destination, due to electrical disturbance an error is introduced in the message. The error detection is used to detect such errors, while error correction technique reconstructs the original data. Parity method is used to detect an error. There are two parity methods to detect an error, namely, odd parity method and even parity method. The extra bit is added to message for detecting a single error, and that extra bit is known as parity bit.

3.7.1 Odd Parity Method

In this method, make number of one's in a message bit odd. For example, a message is 101101, the total number of one's in this message is 4 which is even, add one in the message to make it odd. The extra bit added in the message is known as odd parity bit like 1,101101, the bit before comma is odd parity bit. The odd parity bit is given in Table 3.3 for the numbers from 0 to 9.

Table 3.3 Odd parity bit

Decimal Number	Message (Binary)	Odd Parity Bit
0	0000	1
1	0001	0
2	0010	0
3	0011	1
4	0100	0
5	0101	1
6	0110	1
7	0111	0
8	1000	0
9	1001	1

3.7.2 Even Parity Method

In this method, make number of one's in a message bit even. For example, a message is 101001, the total number of one's in this message is 3 which is odd, add one in the message to make it even. The extra bit added in message is known as even parity bit like 1,101001, the bit before comma is even parity bit. The even parity bit is given in Table 3.4 for the numbers from 0 to 9.

Table 3.4 Even parity bit

Decimal Number	Message (Binary)	Even Parity Bit
0	0000	0
1	0001	1
2	0010	1
3	0011	0
4	0100	1
5	0101	0
6	0110	0
7	0111	1
8	1000	1
9	1001	0

3.7.3 Hamming Code Method

The method is used to detect a single error as well as correct it. You can first find the number of parity bit required to detect and correct the message. To find the number of parity, use the following inequalities equation:

$p + m + 1 \leq 2^p$, where p = number of parity bit and m = number of bits in message.

EXAMPLE 3.24 Find the number of parity bits required to detect and correct the message 1011.

Solution: Here, message = 1011, the total number of bits in the message is 4, so $m = 4$ and $p = ?$ You have to find the value of p.

Put the value of m in the following inequalities:

$$p + m + 1 \leq 2^p$$
$$p + 4 + 1 \leq 2^p$$
$$p + 5 \leq 2^p$$

Put value of $p = 1$

$$1 + 5 \leq 2^1$$

$6 \leq 2$, the value of $p = 1$ is not satisfied the equation, so put value of $p = 2$

$$2 + 4 + 1 \leq 2^2$$
$$7 \leq 2^2$$

$7 \leq 4$, the value of $p = 2$ is also not satisfied, so put $p = 3$

$$3 + 4 + 1 \leq 2^3$$

$8 \leq 8$, now the value of $p = 3$ is satisfying the equation, therefore the total number of three parity bits required to detect as well as correct the given message.

3.7.4 Parity Bit Position

To find the parity bit position, use the following equation:

$$\text{Parity bit position} = 2^n, \quad \text{where } n = 0, 1, 2, 3 \ldots$$

Consider three parity bits p_1, p_2, p_3, the position of the first parity bit $p_1 = 2^0 = 1$. The position of the second parity bit $p_2 = 2^1 = 2$ and the position of the third parity bit $p_3 = 2^2 = 4$.

EXAMPLE 3.25 A message 1011 is transmitted from source to destination. Find the number of parity bits as well as its value using even parity method.

Solution: There are four bits in the given message, so the value of m is 4. Put the value of m in the inequalities $p + m + 1 \leq 2^p$ to find the value of p, where p is the number of bits required to detect as well as correct single bit error in the message.

$$p + 4 + 1 \leq 2^p, \quad p + 5 \leq 2^p$$

Put the value of $p = 1$, the inequalities become $1 + 5 \leq 2^1 = 6 \leq 2$, which is not true.

Put the value of $p = 2$, $2 + 5 \leq 2^2 = 6 \leq 4$, which is again not true.

Put the value of $p = 3$, $3 + 5 \leq 2^3 = 8 \leq 8$, which is now true, so the three numbers of parity bits are required.

Let three parity bits are p_1, p_2 and p_3. Add number of parity bits and number of message bits. The total bits are 7, starting from 1 to 7. The 7 can be represented in three bits. Now find the parity bits position in the seven bit message. The position of parity p_1 is $2^0 = 1$; the first bit will be parity bit p_1 in seven bit message. The position of the second parity bit p_2 is $2^1 = 2$; the second bit of the seven bit message is parity bit p_2. The position of the third parity bit p_3 is $2^2 = 4$; the fourth bit of the seven bit message is third parity bit p_3. The remaining bits in seven messages are given message bit from left to right, see Table 3.5 for more details.

Table 3.5 Bit position

Bit position	1	2	3	4	5	6	7
Binary value	001	010	011	100	101	110	111
Parity bit	p_1	p_2		p_3			
Message bit			1		0	1	1
Parity + message bits	p_1	p_2	1	p_3	0	1	1

Finding the value of p_1

The parity bit p_1 is at bit position one (1), so check the message available at bit position one (1), the bit positions are 1, 3, 5 and 7. Count the number one (1) at this position, if number one is even then the value of p_1 is zero (in the case of even parity method) and one (case of odd parity method). In this example the total number of one's is 2, 2 is even and the even method is used, so the value of $p_1 = 0$.

Finding the value of p_2

The parity bit p_2 is at bit position two (2), so check the message available at bit position two (2), the bit positions are 2, 3, 6 and 7. Count the number one (1) at this position, if the number of one's is even then the value of p_2 is zero (in the case of even parity method) and one (in the case of odd parity method). The total number of message at the given position is three (3) which is odd, so the value of $p_2 = 1$.

Finding the value of p_3

The parity bit p_3 is at bit position three (3), so check the message available at bit position three (3), the bit positions are 4, 5, 6 and 7. Count the number one (1) at this position, if the number of one's is even then the value of p_2 is zero (in the case of even parity method) and one (in the case of odd parity method). The total number of message at the given position is two (2) which is even, so the value of $p_3 = 0$.

The value of parity bit p_1, p_2 and p_3 are 0, 1 and 0, respectively.

3.7.5 Error Position in Message

The message is sent from source to destination; both source and destination are aware about the error detecting parity method used for detecting the error. The source is sending a message using odd parity method, and receiver received the even message means there is an error in the message. But the receiver is not aware about bit position of error. Consider an example; the message 1011 is sent using odd parity method, the number of one's in this message is three (3), which is odd. The receiver received the message 1111 in this message, the total number of one's is four (4) which is even, so there is error in the message. But on which position error is occurred, it is not known to the receiver. The error bit position is detected by calculating the parity bit at source as well as destination. Consider the error vector at source is $v_s = \{sp_1, sp_2, sp_3, ..., sp_n\}$ where $sp_1, sp_2, sp_3, ..., sp_n$ are parity bits at source and vector at destination is $v_d = \{dp_1, dp_2, dp_3, ..., dp_n\}$ where $dp_1, dp_2, dp_3, ..., dp_n$ are parity bit of receiving message. The error position vector is ve = $\{sp_1 \oplus dp_1, sp_2 \oplus dp_2, sp_3 \oplus dp_3, ..., sp_n \oplus dp_n\}$. If ve = 000 means no error in the message. The decimal value of error vector is the error position in the message. Once error position is found, invert the bit to that position to correct the message.

EXAMPLE 3.26 Find the error position as well as correct it in the receiving message 10101101, the odd parity method is used.

Solution: Received message = 1010111, the parity bit in the message at 1, 2 and 4 bit positions, so $sp_1 = 1$, $sp_2 = 0$ and $sp_3 = 0$, this is obtained by formula parity bit position = 2^n where $n = 0, 1, 2, 3...$

$sp_1 = 1$, $sp_2 = 0$ and $sp_3 = 0$, these are the parity bits at the time of sending the message. After discarding the parity bit, the received message is 1111. Now you calculate the destination parity bit as follows:

$m = 4$, now calculate the number of parity bit required to correct the message using formula $p + m + 1 \leq 2^p$

Put the value of $p = 1$, this value does not satisfy the equation, then put the value of $p = 2$ and $p = 3$, $p = 3$ is satisfied the above equation as follows:

$3 + 5 \leq 2^3 = 8 \leq 8$, so there are three parity bits required to correct the message. Let parity bit dp_1, dp_2 and dp_3. The positions of these parity bits in the seven bit message are 2^1, 2^2 and 2^3.

Bit position	1	2	3	4	5	6	7
Binary value	001	010	011	100	101	110	111
Parity bit	dp_1	dp_2		dp_3			
Message bit			1		1	1	1
Parity + message bits	p_1	p_2	1	p_3	1	1	1

Check for dp_1 at 1, 3, 5 and 7. The total number of one's in the message are odd, so $dp_1 = 0$.
Check for dp_2 at 2, 3, 6 and 7. The total number of one's in the message are odd, so $dp_2 = 0$.
Check for dp_3 at 4, 5, 6 and 7. The total number of one's in the message are odd, so $dp_3 = 0$.

Error vector = ve = $\{sp_1 \oplus dp_1, sp_2 \oplus dp_2, sp_3 \oplus dp_3, ..., sp_n \oplus dp_n\}$.

ve = $\{1 \oplus 0, 0 \oplus 0, 0 \oplus 0\} = \{0, 1, 1\}$

ve = 011 = 3, so the error is at third bit position in the message. Revert the number at third bit position to get the correct message. The correct message is 11011.

EXERCISES

Short Answer Questions

1. What do you mean by data representation?
2. What is parity bit?
3. What is odd parity method?
4. What is even parity method?
5. What is Hamming code?
6. Write inequalities for finding the parity bit to detect the error in message.
7. Which method is used to detect single bit error as well as correct it in message?
8. What are BCD and Excess-3 codes?
9. What are the various methods to represent the negative number in computer?
10. How can you find the error position in message?
11. What is Gray code?
12. What is the range of 32-bit register?
13. What is the range of 16-bit register?
14. What is number normalization?
15. Why is number normalization required?

Long Answer Questions

1. Find the 2's complement of the following binary numbers.
   - (i) 1111011
   - (ii) 10001101
   - (iii) 11011.110
   - (iv) 1111011
   - (v) 11111111
2. Find the 1's complement of the following binary numbers.
   - (i) 1110011
   - (ii) 10101101
   - (iii) 11111.110
   - (iv) 1101011
   - (v) 11111011
3. Find the 7's and 8's complements of the following binary numbers.
   - (i) 234
   - (ii) 543
   - (iii) 756
   - (iv) 54324
   - (v) 34567
4. Find the 9's and 10's complements of the following binary numbers.
   - (i) 2348
   - (ii) 9543
   - (iii) 8756
   - (iv) 754324
   - (v) 734567
5. Find the 4's and 5's complements of the following binary numbers.
   - (i) 231
   - (ii) 12233
   - (iii) 3123
   - (iv) 32321
   - (v) 32123

6. Represent the following in 16-bit floating point representation.
   (i) 2.3 (ii) −12.20 (iii) −7
   (iv) −5.5 (v) −3.21
7. Represent the following in 32-bit floating point representation.
   (i) −5.3 (ii) 14.25 (iii) −8
   (iv) −6.9 (v) −7.35
8. What are various methods to represent the negative number?
9. List various codes.
10. Convert the following numbers in BCD, 8421 and 2421 codes.
    (i) 456 (ii) 5432 (iii) 5674
    (iv) 876 (v) 234
11. Represent the following binary numbers into Gray code.
    (i) 11011 (ii) 111001 (iii) 11101101
    (iv) 11110.11011 (v) 11011101
12. Represent the following Gray codes into binary numbers.
    (i) 11111 (ii) 11101 (iii) 110101
    (iv) 1010.11001 (v) 10101101
13. What do you mean by parity bit? Explain odd parity and even parity methods.
14. What do you mean by signed magnitude? Why is signed magnitude used?
16. How many parity bits required for correcting the message 11011?
17. The four parity bit is used to correct the message, then find out the number of bits in message.
18. Write the steps to find the error position in a message.
19. What is Hamming code method? How many errors are detected as well as corrected using this method?
20. Find the value of parity bit for the message 10111 using odd and even methods.
21. The message 1101101 is received then find the parity bit using even parity method.

Multiple Choice Questions

1. The processed data are
   (a) Information
   (b) Data
   (c) Data as well as Information
   (d) None
2. The raw fact is
   (a) Information
   (b) Data
   (c) Information but not Data
   (d) None
3. Which of the following method is used to represent the binary negative number?
   (a) 1's complement
   (b) 2's complement
   (c) Sign Magnitude
   (d) All the above

Data and Information Representation   **63**

4. The 7's complement method is used to represent the negative number in
   (a) Binary Number System   (b) Decimal Number System
   (c) Octal Number System   (d) Hexadecimal Number System
5. The 1's complement of 24 is
   (a) 11000   (b) 00111
   (c) 10011   (d) 00011
6. The 2's complement of 1100111 is
   (a) 0011000   (b) 1100011
   (c) 1100111   (d) 0111011
7. The 8's complement of $(ABC)_{16}$ is
   (a) 110000101011   (b) 101010111100
   (c) 111100011101   (d) 101111000111
8. The 9's complement of 11101 is
   (a) 80   (b) 90
   (c) 70   (d) 60
9. The BCD representation of 45 is
   (a) 01000101   (b) 10101100
   (c) 11011010   (d) 11101101
10. The Gray code representation of 10101 is
    (a) 11111   (b) 10111
    (c) 11011   (d) 11101
11. The binary representation of Gray code 11111 is
    (a) 10101   (b) 11111
    (c) 10001   (d) 10111
12. The decimal number 9 is represented in Excess-3 code is
    (a) 1100   (b) 1110
    (c) 1111   (d) 1011
13. The number of parity bit required to correct the message 1101 is
    (a) 2   (b) 3
    (c) 4   (d) 5
14. If three parity bits are required to detect the error from a message then how many bits are present in the message.
    (a) 3   (b) 5
    (c) 4   (d) 2
15. Which of the following inequality is true to find the number of parity bit required to detect error in message?
    (a) $p + m + 1 \leq 2^p$ (b) $p - m + 1 \leq 2^p$
    (c) $p + m + 1 > 2^p$ (d) $p + m + 1 \geq 2^p$
16. Which of the following formula is valid for finding parity bit position?
    (a) Parity bit position = 2^n where $n = 0, 1, 2, 3...$
    (b) Parity bit position = 2^{n+1} where $n = 0, 1, 2, 3...$

(c) Parity bit position = $2^{n=1}$ where n = 0, 1, 2, 3...
(d) Parity bit position = 2^{n+2} where n = 0, 1, 2, 3...

17. Hamming code method is used to
    (a) Detect a single bit error
    (b) Detect two bits error
    (c) Detect many bits error
    (d) Detect single bit error as well as correct it
18. Parity method is used to
    (a) Detect single bit error only
    (b) Detect two bits and correct it also
    (c) Detect two bits and correct one bit
    (d) None
19. How many types of parity methods
    (a) 1
    (b) 2
    (c) 3
    (d) 4
20. 2's complement of 50 is
    (a) 110010
    (b) 001101
    (c) 001110
    (d) 111101

Answers

1. (a)	2. (b)	3. (d)	4. (c)	5. (b)	6. (a)
7. (b)	8. (c)	9. (a)	10. (a)	11. (a)	12. (a)
13. (b)	14. (c)	15. (a)	16. (a)	17. (d)	18. (a)
19. (b)	20. (c)				

BIBLIOGRAPHY

http://fourier.eng.hmc.edu/e85/lectures/arithmetic_html/node11.html.

Gardner, M.,"The Binary Gray Code", *Knotted Doughnuts and Other Mathematical Entertainments*, W.H. Freeman, New York, 1986, Chapter 2.

Ram, B., *Fundamentals of Micro Computers and Processors*.

CHAPTER 4

Computer Arithmetic

4.1 COMPUTER ARITHMETIC

The computer deals with currents that are on or off, it has been found convenient to represent quantities in binary form to perform arithmetic on a computer. Thus, instead of having ten different digits, 0, 1, 2, 3, 4, 5, 6, 7, 8 and 9, in binary arithmetic, there are only two different digits, 0 and 1, and when moving to the next column, instead of the digit representing a quantity that is ten times as large, it only represents a quantity that is two times as large. We have seen how we can use two digits, 0 and 1, to do the same thing that we can do with the digits 0 and 9 only; write integers equal to or greater than zero. In writing, it is easy enough to add a minus sign to the front of a number, or insert a decimal point. When a number is represented only as a string of bits, with no other symbols, special conventions must be adopted to represent a number that is negative, or one with a radix point indicating a binary fraction.

4.2 BINARY ARITHMETIC

The binary arithmetic means performing the arithmetic on 0 and 1. The addition, subtraction, multiplication and division are performing on 0 and 1; all these operations come under binary arithmetic. In binary arithmetic, carry and borrow are generated at 2, 4, 6 and so on.

4.2.1 Binary Addition

Binary addition means performing arithmetic addition on binary number. The addition of two numbers in base two is known as binary addition. The carry is generated at 2, 4, 6, 8 and so on. The carry generated at 2 is one, at 4 are 2, at 6 are 3 and so on.

EXAMPLE 4.1 Add the two given binary number, $A = 11101$ and $B = 110101$.
Solution: Given $A = 11101$ and $B = 110101$
    The number of digits in A is 5 and 6 in B. So make A six-digit binary number by adding zero before number A as 011101. Now add A and B as follows:

66 Digital Logic Design

Carry	1	2	1	1	0	1	
A		0	1	1	1	0	1
B		1	1	0	1	0	1
Result = Sum	1	0	0	0	0	1	0

Perform addition from left to right, in this addition $1 + 1 = 2$, put 0 as result and use carry as 1, this 1 is added in next number $0 + 0 + 1 = 1$, 1 as sum of next digit, here carry is not generated, do it in this manner to get the result.

4.2.2 Binary Subtraction

Binary subtraction means subtract one binary number from another binary number. The subtraction performed in base two is known as binary subtraction. In this subtraction borrow is taken as 2, like 10 in decimal number.

One's complement subtraction

The binary subtraction is performing using one's complement is known as one's complement subtraction. The algorithm for subtraction using one's complement method is as follows:

Step 1: Convert the given number in binary number; use 5 or more binary digit for representing number. For example, 1011, represent 1011 in five digits as 01011.
Step 2: If the number of digits in both numbers is not equal, then make number of digits equal by adding zero before the number which digits are less.
Step 3: Find one's complement of negative number.
Step 4: Add one's complement obtained for negative number in step 2 to given positive number.
Step 5: If carry is generated in step 4, then add carry to the sum obtained in step 4 from final result, else the sum obtained in step 4 is final result.

Note: If carry generated in step 5 means result is positive else result is negative.

EXAMPLE 4.2 Perform 22–45 subtraction using one's complement.

Solution: Let $A = 22$ and $B = 45$. Now convert A and B in binary numbers.

Step 1: $A = 22 = 16 + 8 + 4 + 2 + 1 = 1 + 0 + 1 + 1 + 0 = 10110$
$B = 45 = 32 + 16 + 8 + 4 + 2 + 1 = (32 + 8 + 4 + 1 = 45)$
$= 1 + 0 + 1 + 1 + 0 + 1 = 101101$

Step 2: Here A is of 5-bit number and B is of 6-digit number, so add zero before A to make 6 digits equal to the number of digit in B, $A = 010110$. Now $A = 010110$, $B = 101101$

Step 3: Find one's complement of B
Let 1's complement of B is $B' = 111111 - B = 111111 - 101101 = 010010$.

Step 4: Add $A + B'$ (which is complement of B)

```
  010110
  010010
  ──────
  101000
```

Here carry is not generated, so the result is 101000.

Check the answer by direct method: $22 - 45 = -23$ = binary number of 23, 6 digit = $32 + 16 + 8 + 4 + 2 + 1 = 010111$, one's complement of it is $-23 = 101000$; this result is same as the result obtained in step 4. Therefore, obtained result is correct.

EXAMPLE 4.3 Perform $45 - 22$ subtraction using one's complement.

Solution:

Step 1: $A = 45 = 32 + 16 + 8 + 4 + 2 + 1 = (32 + 8 + 4 + 1 = 45)$
$$= 1 + 0 + 1 + 1 + 0 + 1 = 101101$$
$B = 22 = 16 + 8 + 4 + 2 + 1 = 1 + 0 + 1 + 1 + 0 = 10110$

Step 2: Here A is of 6 bits number and B is of 5 digits number, so add zero before B to make 6 digits equal to the number of digits in $A, B = 010110$.
Now $A = 101101$, $B = 010110$

Step 3: Find one's complement of B
Let 1's complement of B is $B' = 111111 - B = 111111 - 010110 = 101001$.

Step 4: Add $A + B'$ (which is complement of B)

$$\begin{array}{r} 101101 \\ 101001 \\ \hline \end{array}$$

(Carry) 1,010110 (Result)

Here carry is generated, so add this carry to the result like as follows:
Final result = $010110 + 1 = 010111$
The final result is 010111.

Note: Here carry is generated, so the result is positive.

Check the answer by direct method: $45 - 22 = 23$ = binary number of 23, 6 digits = $32 + 16 + 8 + 4 + 2 + 1 = 010111$. Here no need to find one's complement because number is in positive form. This result is same as the result obtained in step 4. Therefore, obtained result is correct.

Two' complement subtraction

The binary subtraction is performing using 2's complement is known as 2's complement subtraction. The algorithm for subtraction using 2's complement method is as follows:

Step 1: Convert the given number in binary number; use 5 or more binary digits for representing the number. For example, 1011, represent 1011 in five digits as 01011.

Step 2: If number of digits in both numbers are not equal, then make number of digits equal by adding zero before the number which digits are less.

Step 3: Find 2's complement of negative number.

Step 4: Add 2's complement obtained for negative number in step 2 to given positive number.

Step 5: If carry is generated in step 4, ignore it from sum obtained in step 4 is the final result.

Note: If carry is generated in step 5 means result is positive else result is negative.

EXAMPLE 4.4 Perform 22 − 45 subtraction using 2's complement.

Solution: Let $A = 22$ and $B = 45$. Now convert A and B in binary number

Step 1: $A = 22 = 16 + 8 + 4 + 2 + 1 = 1 + 0 + 1 + 1 + 0 = 10110$
$B = 45 = 32 + 16 + 8 + 4 + 2 + 1 = (32 + 8 + 4 + 1 = 45)$
$= 1 + 0 + 1 + 1 + 0 + 1 = 101101$

Step 2: Here A is of 5 bits number and B is of 6 digits number, so add zero before A to make 6 digits equal to the number of digit in B, $A = 010110$.
Now $A = A = 010110$, $B = 101101$.

Step 3: Find 2's complement of B
Let 2's complement of B is
$B' = 111111 − B + 1 = 111111 − 101101 + 1 = 010010 + 1 = 010011$.

Step 4: Add $A + B'$ (which is 2's complement of B)

$$\begin{array}{r} 010110 \\ \underline{010011} \\ 101001 \end{array}$$

Here carry is not generated, so the result is 101001.

Check the answer by direct method: $22 − 45 = −23$ = binary number of 23, 6 digits = $32 + 16 + 8 + 4 + 2 + 1 = 010111$, 2's complement of it is $−23 = 101001$; this result is same as the result obtained in step 4. Therefore, obtained result is correct.

EXAMPLE 4.5 Perform 45 − 22 subtraction using 2's complement.

Solution:

Step 1: $A = 45 = 32 + 16 + 8 + 4 + 2 + 1 = (32 + 8 + 4 + 1 = 45)$
$= 1 + 0 + 1 + 1 + 0 + 1 = 101101$
$B = 22 = 16 + 8 + 4 + 2 + 1 = 1 + 0 + 1 + 1 + 0 = 10110$

Step 2: Here A is of 6 bits number and B is of 5 digits number, so add zero before B to make 6 digits equal to the number of digit in A, $B = 010110$.
Now $A = 101101$, $B = 010110$

Step 3: Find 2's complement of B
Let 2's complement of B is $B' = 111111 − B + 1 = 111111 − 010110 + 1 = 101010$.

Step 4: Add $A + B'$ (which is 2's complement of B)

$$\begin{array}{r} 101101 \\ \underline{101010} \end{array}$$

(Carry) 1,010111 (Result)

Here carry is generated, ignore it, sum after ignoring carry is the final result. The final result is 010111.

Note: Here carry is generated, so the result is positive.

Check the answer by direct method: $45 - 22 = 23$ = binary number of 23, 6 digits = $32 + 16 + 8 + 4 + 2 + 1 = 010111$. Here no need to find 2's complement because number is in positive. This result is same as the result obtained in step 4. Therefore, obtained result is correct.

4.2.3 Binary Multiplication

Binary multiplication is performed same as decimal number multiplication but base two (2) is considered in binary multiplication while in decimal number multiplication base is considered as ten (10). The rule for binary multiplication is as follows:

$$0 \times 0 = 0$$
$$0 \times 1 = 0$$
$$1 \times 0 = 0$$
$$1 \times 1 = 1$$

EXAMPLE 4.6 Perform the following binary multiplication

$$1011 \times 11$$

Solution:

```
      1 0 1 1
    ×     1 1
    ─────────
      1 0 1 1
    1 0 1 1
    ─────────
    1 0 0 0 0 1
```

EXAMPLE 4.7 If $m = 11101$ and $n = 1101$. Find $m \times n$.

Solution:

$$m \times n$$

```
Carry =   1 2 2 2 1 1 1
                1 1 1 1 0 1
              ×     1 1 0 1
              ─────────────
                1 1 1 1 0 1
              0 0 0 0 0 0
            1 1 1 1 0 1
          1 1 1 1 0 1
          ───────────────
          1 0 0 0 1 1 0 0 1
```

4.2.4 Binary Division

Binary addition is also same as decimal division, difference is simply in base. Base two is used in binary division while base ten is used in decimal division.

EXAMPLE 4.8 Perform the following binary division

$$1101 \div 11$$

Solution:

$$11 \overline{)1101}(1.01$$
$$\underline{11}$$
$$00100$$
$$\underline{11}$$
$$1$$

4.3 OCTAL ARITHMETIC

The arithmetic is performed on octal number is known as octal arithmetic. In this type of arithmetic the base is consider as eight.

4.3.1 Octal Addition

Octal addition means performing arithmetic addition on octal number. The addition of two numbers in base eight is known as octal addition. The carry is generated at 8, 16, 24, 32 and so on. The carry generated at 8 is one, at 16 are 2, at 24 are 3 and so on.

EXAMPLE 4.9 Add the two given octal numbers, $A = 567$ and $B = 675$.

Solution: Given $A = 567$ and $B = 675$
Now add A and B as follows:

Carry		1	1	1
A			5	6	7
B			6	7	5
Result = Sum		1	4	6	4

Perform addition from left to right, in this addition 7 + 5 = 12, here 12 = 8 + 4, so write 4 as sum and carry one, in sum of next digit (carry) 1 + 6 + 7 = 14, here 14 = 8 + 6, write 6 as sum and again carry one, do in this manner until end of last digit of the given number.

4.3.2 Octal Subtraction

Octal subtraction means subtract one octal number from another octal number. The subtraction performed in base eight is known as octal subtraction. In this subtraction borrow is taken as 8, like 10 in decimal number.

Seven's complement subtraction

The octal subtraction is performing by using 7's complement is known as 7's complement subtraction. The algorithm for subtraction by using 7's complement method is as follows:

Step 1: Convert the given number in octal number, if the number is not given in octal number.
Step 2: If number of digits in both numbers is not equal, then make number of digits equal by adding zero before the number which digits are less.
Step 3: Find 7's complement of negative number.
Step 4: Add 7's complement obtained for negative number in step 2 to given positive number.
Step 5: If carry is generated in step 4, ignore it from sum obtained in step 4 and add one to the sum to get the final result.

Note: If carry generated in step 4 means result is positive else result is negative.

EXAMPLE 4.10 Perform $622 - 245$ subtraction by using 7's complement, number is given in octal form.

Solution: Let $A = 622$ and $B = 245$.
$$A = 622$$
$$B = -245$$
Let 7's complement of B is $B' = 777 - B = 777 - 245 = 532$
Now, add $A + B'$ (which is 7's complement of B)

$$\begin{array}{r} 622 \\ 532 \\ \hline \text{Carry(1)} \; 354 \end{array}$$

Here carry is generated, so the result is $354 + 1 = 355$.

Check the answer by direct method: Here octal subtraction is $622 - 245 = 355$; this result is same as the result obtained in step 4. Therefore, obtained result is correct.

EXAMPLE 4.11 Perform $122 - 245$ subtraction by using 7's complement, number is given in octal form.

Solution: Let $A = 122$ and $B = 245$
$$A = 122$$
$$B = -245$$
Let 7's complement of B is $B' = 777 - B = 777 - 245 = 532$
Now, add $A + B'$ (which is 7's complement of B)

$$\begin{array}{r} 122 \\ 532 \\ \hline 654 \end{array}$$

Here carry is not generated, so the result is negative, the result is 654.

Check the answer by direct method: Here octal subtraction is $622 - 245 = -355$; this result is same as result obtained in step 4. Therefore, obtained result is correct.

Eight's complement subtraction

The octal subtraction performed by using 8's complement is known as 8's complement subtraction. The algorithm for subtraction by using 8's complement method is as follows:

Step 1: Convert the given number in octal form, if the number is not given in octal form.
Step 2: If the number of digits in both numbers is not equal, then make the number of digits equal by adding zero before the number which digits are less.
Step 3: Find 8's complement of negative number.
Step 4: Add 8's complement obtained for negative number in step 2 to given positive number.
Step 5: If carry is generated in step 4, ignore it from sum obtained in step 4 is final result.

Note: If carry generated in step 5 means result is positive else result is negative.

EXAMPLE 4.12 Perform 645 – 22 subtraction by using 8's complement (number is in octal form).

Solution:
Step 1: $A = 645$
$B = 422$
Step 2I: Here A is of 3-digit number and B is of 2-digit number, so add zero before B to make 3 digits equal to the number of digit in A, $B = 022$.
Now $A = 645$, $B = 022$
Step 3: Find 8's complement of B
Let 8's complement of B is $B' = 777 - B + 1 = 777 - 022 + 1 = 755 + 1 = 756$
Step 4: Add $A + B'$ (which is 8's complement of B)

```
        645
        756
      _____
(Carry) 1,623  (Result)
```

Here carry is generated, ignore it, sum after ignoring carry is the final result.
The final result is 623.

Check the answer by direct method: $645 - 022 = 623$. This result is same as the result obtained in step 4. Therefore, obtained result is correct.

EXAMPLE 4.13 Perform 45 – 122 subtraction using 8's complement (number is in octal form).

Solution:
Step 1: $A = 45$
$B = 122$
Step 2: Here A is of 2-digit number and B is of 3-digit number, so add zero before A to make 3 digits equal to the number of digit in B, $A = 045$.
Now $A = 045$, $B = 122$
Step 3: Find 8's complement of B. Let 8's complement of B is
$B' = 777 - B + 1 = 777 - 122 + 1 = 655 + 1 = 656$
Step 4: Add $A + B'$ (which is 8's complement of B)

```
   045
   656
  _____
   723
```

Here carry is not generated, so the result is negative. The final result is 723.

Computer Arithmetic   **73**

Check the answer by direct method: 045 − 122 = −65. The 8's complement of this number is 777 − 065 + 1 = 712 + 1 = 713; this result is same as the result obtained in step 4. Therefore, obtained result is correct.

4.4   DECIMAL ARITHMETIC

Decimal arithmetic is used in daily life. Decimal addition is well known by everyone. This is generally used in daily life.

9's complement subtraction

The decimal subtraction performed by using 9's complement is known as 9's complement subtraction. The algorithm for subtraction by using 9's complement method is as follows:

Step 1: Convert the given number in decimal form, if the number is not given in decimal form.
Step 2: If the number of digits in both numbers is not equal, then make number of digits equal by adding zero before the number which digits are less.
Step 3: Find 9's complement of negative number.
Step 4: Add 9's complement obtained for negative number in step 2 to given positive number.
Step 5: If carry is generated in step 4, add this carry to sum obtained in step 4 which is the final result.

Note: If carry generated in step 5 means result is positive else result is negative.

EXAMPLE 4.14   Perform 622 − 245 subtraction by using 9's complement, if the number is given in decimal form.

Solution: Let $A = 622$ and $B = 245$.
$A = 622$
$B = -245$
Let 9's complement of B is $B' = 999 − B = 999 − 245 = 754$
Now, add $A + B'$ (which is 9's complement of B)

$$\begin{array}{r} 622 \\ 754 \\ \hline \text{Carry}(1)\ \ 376 \end{array}$$

Here carry is generated, so the result is 376 + 1 = 377

Check the answer by direct method: Here decimal subtraction is 622 − 245 = 377; this result is same as the result obtained in step 4. Therefore, obtained result is correct.

EXAMPLE 4.15   Perform 122 − 245 subtraction by using 9's complement, if the number is given in decimal form.

Solution: Let $A = 122$ and $B = 245$.

$$A = 122$$
$$B = -245$$

Let 9's complement of B is $B' = 999 - B = 999 - 245 = 754$
Now, adds $A + B'$ (which is 9's complement of B)

```
  122
  754
  ---
  876
```

Here carry is not generated, so the result is 876.

Check the answer by direct method: Here decimal subtraction is $122 - 245 = -123$, 9's complement of this number is $999 - 123 = 876$; this result is same as the result obtained in step 4. Therefore, obtained result is correct.

10's complement subtraction

The decimal subtraction performed by using 10's complement is known as 10's complement subtraction. The algorithm for subtraction by using 10's complement method is as follows:

Step 1: Convert the given number in decimal form, if the number is not given in decimal form.
Step 2: If the number of digits in both numbers is not equal, then make number of digits equal by adding zero before the number which digits are less.
Step 3: Find 10's complement of negative number.
Step 4: Add 10's complement obtained for negative number in step 2 to given positive number.
Step 5: If carry is generated in step 4, add this carry to sum obtained in step 4 which is the final result.

Note: If carry generated in step 5 means result is positive else result is negative.

EXAMPLE 4.16 Perform $645 - 22$ subtraction by using 10's complement (number is in decimal form).

Solution:
Step 1: $A = 645$
$B = 22$
Step 2: Here A is of 3-digit number and B is of 2-digit number, so add zero before B to make 3 digits equal to the number of digit in A, $B = 022$.
Now $A = 645$, $B = 022$
Step 3: Find 10's complement of B. Let 10's complement of B is $B' = 999 - B + 1 = 999 - 022 + 1 = 977 + 1 = 978$
Step 4: Add $A + B'$ (which is 10's complement of B)

```
           645
           978
           ---
 (Carry)  1,623  (Result)
```

Here carry is generated, ignore it, sum after ignoring carry is the final result. The final result is 623.

Check the answer by direct method: 645 – 022 = 623. This result is same as the result obtained in step 4. Therefore, obtained result is correct.

EXAMPLE 4.17 Perform 134 – 322 subtraction using 10's complement (number is in decimal).
Solution:
Step 1: $A = 134$
          $B = 322$
Step 2:   Here A is of 3-digit number and B is of 3-digit number, both numbers are of equal number of digits, so no need to add zero to any number.
Step 3:   Find 10's complement of B
          Let 10's complement of B is $B' = 999 - B + 1 = 999 - 322 + 1 = 677 + 1 = 678$
Step 4:   Add $A + B'$ (which is 10's complement of B)

          134
          678
          ―――
          812   (Result)

Here carry is not generated, so the result is in negative, the final result is 812.

Check the answer by direct method: 134 – 322 = –188. The 10's complement of 188 is 999 – 188 + 1 = 811 + 1 = 812; this result is same as the result obtained in step 4. Therefore, obtained result is correct.

4.5   HEXADECIMAL ARITHMETIC

The hexadecimal arithmetic means addition, subtraction, multiplication and division are performed on base sixteen.

4.5.1   Addition

Hexadecimal addition means performing arithmetic addition on hexadecimal number. The addition of two numbers in base sixteen is known as hexadecimal addition. The carry is generated at 16, 32, 48, 64 and so on. The carry generated at 16 is one, at 32 are 2, at 48 are 3 and so on.

EXAMPLE 4.18   Add the two given hexadecimal numbers, $A = 56C$ and $B = 67D$.
Solution: Given $A = 567$ and $B = 675$
   Now add A and B as follows:

Carry		1	
A	5	6	C
B	6	7	D
Result = Sum	B	E	9

Perform addition from left to right, in this addition C(12) + D(13) = 25 = 25 − 16 = 9, here 9 is sum of last two digits of the given number and carry one, the sum of next digits (carry)1 + 6 + 7 = 14 = E, after next digit there is no carry, do in this manner until end of last digit of the given number.

4.5.2 Hexadecimal Subtraction

Hexadecimal subtraction means subtract one hexadecimal number from another hexadecimal number. The subtraction performed in base sixteen is known as hexadecimal subtraction. In this subtraction borrow is taken as 16, like 10 in decimal number.

15's complement subtraction

The hexadecimal subtraction performed by using 15's complement is known as 15's complement subtraction. The algorithm for subtraction by using 15's complement method is as follows:

Step 1: Convert the given number in hexadecimal number, if the number is not given in hexadecimal form.
Step 2: If the number of digits in both numbers is not equal, then make number of digits equal by adding zero before the number which digits are less.
Step 3: Find 15's complement of negative number.
Step 4: Add 15's complement obtained for negative number in step 2 to given positive number.
Step 5: If carry is generated in step 4, add it to the sum obtained in step 4 which is the final result.

Note: If carry generated in step 5 means result is positive else result is negative.

EXAMPLE 4.19 Perform 6B2 − 2A5 subtraction by using 15's complement; if the number is given in hexadecimal form.

Solution: Let x = 6B2 and y = 2A5.
Let 15's complement of y is y' = FFF(15 15 15) − x = FFF(15 15 15) − 2A5 = D5A
Now, add $x + y'$ (which is 15's complement of y)

$$\begin{array}{r} 6\ B\ 2 \\ D\ 5\ A \\ \hline \text{Carry(1)}\ \ 4\ 0\ C \end{array}$$

Here carry is generated, so the result is 40C + 1 = 40D.

Check the answer by direct method: Here hexadecimal subtraction is 6B2 − 2A5 = 40D, this result is same as the result obtained in step 4. Therefore, obtained result is correct.

EXAMPLE 4.20 Perform 12A − 2B5 subtraction by using 15's complement, if the number is given in hexadecimal form.

Solution: Let x = 12A and y = 2B5.
x = 12A

Computer Arithmetic   **77**

$$y = -2B5$$

Let 15's complement of y is $y' = $ FFF(15 15 15) $- y = $ FFF(15 15 15) $-$ 2B5 = D4A
Now, add $x + y'$ (which is 15's complement of y)

```
  1 2 A
  D 4 A
  -----
  E 7 4
```

Here carry is not generated, so the result is negative, the result is E74.

Check the answer by direct method: Here hexadecimal subtraction is 12A $-$ 2B5 = $-$75, the 15's complement of this number is FFF(15 15 15) $-$ 75 = E6A; this result is same as the result obtained in step 4. Therefore, obtained result is correct.

16's complement subtraction

The hexadecimal subtraction performed by using 16's complement is known as 16's complement subtraction. The algorithm for subtraction by using 16's complement method is as follows:

Step 1: Convert the given number in hexadecimal form, if the number is not given in hexadecimal form.
Step 2: If the number of digits in both numbers is not equal, then make number of digits equal by adding zero before the number which digits are less.
Step 3: Find 16's complement of negative number.
Step 4: Add 16's complement obtained for negative number in step 2 to given positive number.
Step 5: If carry is generated in step 4, add it to the sum obtained in step 4 which is the final result.

Note: If carry generated in step 5 means result is positive else result is negative.

EXAMPLE 4.21 Perform 645 $-$ 22 subtraction by using 16's complement (number is in hexadecimal form).

Solution:
Step 1: $x = 645$
$y = 22$
Step 2: Here x is of 3-digit numbers and y is of 2-digit numbers, so add zero before y to make 3-digits equal to the number of digits in x, $y = 022$.
Now $x = 645$, $y = 022$
Step 3: Find 16's complement of x
Let 16's complement of y is $y' = $ FFF(15 15 15) $- y + 1$
= FFF(15 15 15) $-$ 022 + 1 = FDD + 1 = FDE
Step 4: Add $x + y'$ (which is 16's complement of y)

```
         6 4 5
         F D E
         -----
(Carry) 1, 6 2 3  (Result)
```

78 Digital Logic Design

Here carry is generated, ignore it, sum after ignoring carry is the final result. The final result is 623.

Check the answer by direct method: 645 – 022 = 623. This result is same as the result obtained in step 4. Therefore, obtained result is correct.

EXAMPLE 4.22 Perform 45 – 122 subtraction by using 16's complement (number is in hexadecimal form).

Solution:
Step 1: $x = 45$
         $y = 122$
Step 2: Here x is of 2-digit number and y is of 3-digit number, so add zero before x to make 3 digits equal to the number of digits in y, $x = 045$.
         Now $x = 045$, $y = 122$
Step 3: Find 16's complement of x
         Let 16's complement of y is $y' =$ FFF(15 15 15) – y + 1
         = FFF(15 15 15) – 122 + 1 = EDD + 1 = EDE
Step 4: Add $x + y'$ (which is 16's complement of y)

```
  0 4 5
  E D E
  -----
  F 1 3
```

Here carry is not generated, so the result is negative. The final result is F13.

Check the answer by direct method: 045 – 122 = –123. The 16's complement of this number is FFF(15 15 15) – 123 + 1 = EDC + 1 = EDD; this result is same as the result obtained in step 4. Therefore, obtained result is correct.

4.6 HEXAL ARITHMETIC

Hexal arithmetic means performing addition, subtraction, multiplication and division on hexal number. The addition, subtraction, multiplication and division are performing on base six. The carry is generated at 6, 12, 18, 24 and so on. The carry generated at 6 is one, at 12 are 2, and at 18 are 3 and so on.

4.6.1 Hexal Addition

Hexal addition means performing addition of two or more hexal numbers. The example of hexal addition is given below.

EXAMPLE 4.23 Add the two given hexal numbers, $A = 134$ and $B = 542$.

Solution: Given $A = 134$ and $B = 542$.
   Now add A and B as follows:

Computer Arithmetic

Carry		1	1	
A		1	3	4
B		5	4	2
Result = Sum	1	0	2	0

Perform addition from left to right, in this addition 4 + 2 = 6 = 6 – 6 = 0, here 0 is sum of last two digits of the given number and carry one, the sum of next digits is (carry) 1 + 3 + 4 = 7 = 7 – 6 = 1 and carry again one, after next digits there is one carry, do in this manner until end of last digit of the given number.

4.6.2 Hexal Subtraction

Hexal subtraction means subtract one hexal number from another hexal number. The subtraction performed in base six is known as hexal subtraction. In this subtraction borrow is taken as 6, like 10 in decimal number.

EXAMPLE 4.24 Perform the hexal subtraction $A - B$ of two given hexal numbers, $A = 334$ and $B = 142$.

Solution: Given $A = 334$ and $B = 142$.

Now perform $A - B$ as follows:

Carry		1	
A	3	3	4
B	1	4	2
Result = difference	1	5	2

Perform subtraction from left to right, in this subtraction 4 – 2 = 2, here 2 is difference of last two digits of the given number and borrow is zero because A's digit is greater than B's digit, the difference of next digit (borrow 1 means add 6 in 3 once borrow is taken from previous number, the previous number is reduce by one) like 6 + 3 – 4 = 5 is difference of next digit, do in this manner till last digit of the number.

4.6.3 Hexal Multiplication

Hexal multiplication means multiply two hexal numbers in base six. The multiplication performed in base six is known as hexal multiplication. In this multiplication carry is taken on 6, like 10 in decimal number.

EXAMPLE 4.25 Perform hexal multiplication $A \times B$ of two given hexal numbers, $A = 334$ and $B = 142$.

Solution: Given $A = 334$ and $B = 142$.

Now perform $A \times B$ as follows:

Carry				2, 1	0,2,1	0,2,1
A				3	3	4
B				1	4	2
Multiplication by 2 of A to each digit of B			1	1	1	2
Shift one digit left and multiply 4 of A to each digit of B		2	2	2	4	
Shift two digits left and multiply 1 of A to each digit of B		3	3	4		
Sum of multiplication = Result = Multiplication of $A \times B$	1	0	1	1	5	2

Multiplication algorithm

Step 1: Let A multiplicand and B multiplier

Step 2: Multiply each digit of A by last digit of B and record it if carry is generated, store it as carry, and add it to next digit of A multiplication and do it until the end of digit of A.

Step 3: Shift one digit from left to right and take next digit of B, do as in step 2.

Step 4: Do step 2 to step 3 until you reach to the first digit of B.

Step 5: Perform hexal addition for all multiplication to get the result.

4.6.4 Hexal Division

Hexal division means divide one hexal decimal number by another hexal decimal number in base six. The division performed in base six is known as hexal division. In this division carry is taken on 6, like 10 in decimal number.

15's complement subtraction

The hexadecimal subtraction performed by using 15's complement is known as 15's complement subtraction. The algorithm for subtraction by using 15's complement method is as follows:

Step 1: Convert the given number in hexadecimal form, if the number is not given in hexadecimal form.

Step 2: If the number of digits in both numbers is not equal, then make the number of digits equal by adding zero before the number which digits are less.

Step 3: Find 15's complement of negative number.

Step 4: Add 15's complement obtained for negative number in step 2 to given positive number.

Step 5: If carry is generated in step 4, add it to the sum obtained in step 4 which is the final result.

Note: If carry generated in step 5 means result is positive else result is negative.

EXAMPLE 4.26 Perform 6B2 – 2A5 subtraction by using 15's complement; if the number is given in hexadecimal form.

Solution: Let x = 6B2 and y = 2A5.

Let 15's complement of y is y' = FFF(15 15 15) – x = FFF(15 15 15) – 2A5 = D5A

Now, add $x + y'$ (which is 15's complement of y)

```
            6 B 2
            D 5 A
 Carry(1)   4 0 C
```

Here carry is generated, so the result is 40C + 1 = 40D.

Check the answer by direct method: Here hexadecimal subtraction is 6B2 − 2A5 = 40D; this result is same as the result obtained in step 4. Therefore, obtained result is correct.

EXAMPLE 4.27 Perform 12A − 2B5 subtraction by using 15's complement, if the number is given in hexadecimal form.

Solution: Let $\qquad x = 12A$ and $y = 2B5$.
$\qquad\qquad\qquad\qquad x = 12A$
$\qquad\qquad\qquad\qquad y = -2B5$

Let 15's complement of y is $y' = $ FFF(15 15 15) − y = FFF(15 15 15) − 2B5 = D4A
Now, add $x + y'$ (which is 15's complement of y)

```
    1 2 A
    D 4 A
    E 7 4
```

Here carry is not generated, so the result is negative, the result is E74.

Check the answer by direct method: Here hexadecimal subtraction is 12A − 2B5 = −75, the 15's complement of this number is FFF(15 15 15) − 75 = E6A; this result is same as the result obtained in step 4. Therefore, obtained result is correct.

16's complement subtraction

The hexadecimal subtraction performed by using 16's complement is known as 16's complement subtraction. The algorithm for subtraction by using 16's complement method is as follows:

Step 1: Convert the given number in hexadecimal form, if the number is not given in hexadecimal form.
Step 2: If the number of digits in both numbers is not equal, then make the number of digits equal by adding zero before the number which digits are less.
Step 3: Find 16's complement of negative number.
Step 4: Add 16's complement obtained for negative number in step 2 to given positive number.
Step 5: If carry is generated in step 4, add it to the sum obtained in step 4 which is the final result.

Note: If carry generated in step 5 means result is positive else result is negative.

EXAMPLE 4.28 Perform 645 − 22 subtraction by using 16's complement (number is in hexadecimal).

82 Digital Logic Design

Solution:
Step 1: $x = 645$
$y = 22$
Step 2: Here x is of 3-digit number and y is of 2-digit number, so add zero before y to make 3 digits equal to the number of digit in x, $y = 022$.
Now $x = 645$, $y = 022$
Step 3: Find 16's complement of x
Let 16's complement of y is $y' = $ FFF(15 15 15) $- y + 1$
$= $ FFF(15 15 15) $- 022 + 1 = $ FDD $+ 1 = $ FDE
Step 4: Add $x + y'$ (which is 16's complement of y)

$$\begin{array}{r} 6\ 4\ 5 \\ \text{F D E} \\ \hline \end{array}$$
(Carry) 1, 6 2 3 (Result)

Here carry is generated, ignore it, sum after ignoring carry is the final result.
The final result is 623.

Check the answer by direct method: $645 - 022 = 623$. This result is same as the result obtained in step 4. Therefore, obtained result is correct.

EXAMPLE 4.29 Perform $45 - 122$ subtraction by using 16's complement (number is in hexadecimal).

Solution:
Step 1: $x = 45$
$y = 122$
Step 2: Here x is of 2-digit number and y is of 3-digit number, so add zero before x to make 3 digits equal to the number of digit in y, $x = 045$.
Now $x = 045$, $y = 122$
Step 3: Find 16's complement of x
Let 16's complement of y is $y' = $ FFF(15 15 15) $- y + 1$
$= $ FFF (15 15 15) $- 122 + 1 = $ EDD $+ 1 = $ EDE
Step 4: Add $x + y'$ (which is 16's complement of y)

$$\begin{array}{r} 0\ 4\ 5 \\ \text{E D E} \\ \hline \text{F 1 3} \end{array}$$

Here carry is not generated, so the result is negative. The final result is F13.

Check the answer by direct method: $045 - 122 = -123$. The 16's complement of this number is FFF(15 15 15) $- 123 + 1 = $ EDC $+ 1 = $ EDD; this result is same as the result obtained in step 4. Therefore, obtained result is correct.

4.7 NONAL ARITHMETIC

Nonal arithmetic means performing addition, subtraction, multiplication and division on nonal number. The addition, subtraction, multiplication and division are performing on base nine. The carry is generated at 9, 18, 27, and so on. The carry generated at 9 is one, at 18 are 2, and at 27 are 3 and so on.

4.7.1 Nonal Addition

Nonal addition means performing addition of two or more nonal numbers. The example of nonal addition is given below.

EXAMPLE 4.30 Add the two given nonal numbers, $A = 138$ and $B = 547$.
Solution: Given $A = 138$ and $B = 547$.
 Now add A and B as follows:

Carry	0	1	
A	1	3	8
B	5	4	7
Result = Sum	6	8	6

Perform addition from left to right, in this addition $8 + 7 = 15 - 9 = 6$, here 6 is sum of last two digits of the given number and carry one, the sum of next digits (carry) $1 + 3 + 4 = 8$, 8 is less than 8, so you cannot subtract nine as previous digit. Here carry is zero, add zero to the sum of next digit, and do in this manner until end of last digit of the given number.

4.7.2 Nonal Subtraction

Nonal subtraction means subtract one nonal number from another nonal number. The subtraction performed in base nine is known as nonal subtraction. In this subtraction borrow is taken as 9, like 10 in decimal number.

EXAMPLE 4.31 Perform the nonal subtraction $A - B$ of two given nonal numbers, $A = 334$ and $B = 142$.
Solution: Given $A = 334$ and $B = 142$.
 Now perform $A - B$ as follows:

Carry		1	
A	3	3	4
B	1	4	2
Result = difference	1	8	2

Perform subtraction from left to right, in this subtraction $4 - 2 = 2$, here 2 is difference of last two digits of the given number and borrow is zero because A's digit is greater than

B's digit, the difference of next digits (borrow 1 means add 9 in 3 once borrow is taken from previous number, the previous number is reduced by one) like 9 + 3 – 4 = 8 is difference of next digits, do in this manner till last digit of the number.

8's complement subtraction (nonal number)

The nonal number subtraction performed by using 8's complement is known as 8's complement subtraction. The algorithm for performing subtraction nonal number by using 8's complement method is as follows:

Step 1: Convert the given number in nonal form, if the number is not given in nonal form.
Step 2: If the number of digits in both numbers is not equal, then make number of digits equal by adding zero before the number which digits are less.
Step 3: Find 8's complement of negative number.
Step 4: Add 8's complement obtained for negative number in step 2 to given positive number.
Step 5: If carry is generated in step 4, add it to the sum obtained in step 4 which is the final result.

Note: If carry generated in step 5 means result is positive else result is negative.

EXAMPLE 4.32 Perform 682 – 275 subtraction by using 8's complement, if the number is given in nonal form.

Solution: Let x = 682 and y = 275.
Let 8's complement of y is y' = 888 – y = 888 – 275 = 613
Now, add $x + y'$ (which is 8's complement of y)

```
           682
           613
-----------------
Carry(1)   405
```

Here carry is generated, so the result is 405 + 1 = 406

Check the answer by direct method: Here nonal subtraction is 682 – 275 = 406, this result is same as the result obtained in step 4. Therefore, obtained result is correct.

EXAMPLE 4.33 Perform 128 – 287 subtraction by using 8's complement, if the number is given in nonal form.

Solution: Let x = 128 and y = 287.
x = 128
y = –287
Let 8's complement of y is y' = 888 – y = 888 – 287 = 601
Now, add $x + y'$ (which is 8's complement of y)

```
1 2 8
6 0 1
-------
7 3 0
```

Here carry is not generated, so the result is negative, the result is 730.

Computer Arithmetic **85**

Check the answer by direct method: Here nonal number subtraction is 128 − 287 = −158, the 8's complement of this number is 888 − 158 = 730; this result is same as the result obtained in step 4. Therefore, obtained result is correct.

9's complement subtraction (nonal number)

The nonal subtraction performed by using 9's complement is known as 9's complement subtraction. The algorithm for subtraction of nonal number by using 9's complement method is as follows:

Step 1: Convert the given number in nonal form, if the number is not given in nonal form.
Step 2: If the number of digits in both numbers is not equal, then make number of digits equal by adding zero before the number which digits are less.
Step 3: Find 9's complement of negative number.
Step 4: Add 9's complement obtained for negative number in step 2 to given positive number.
Step 5: If carry is generated in step 4, add it to the sum obtained in step 4 which is the final result.

Note: If carry generated in step 5 means result is positive else result is negative.

EXAMPLE 4.34 Perform 645 − 22 subtraction by using 9's complement (number is in nonal form).

Solution:
Step 1: $x = 645$
$y = 22$
Step 2: Here x is of 3-digit number and y is of 2-digit number, so add zero before y to make 3 digits equal to the number of digit in x, $y = 022$.
Now $x = 645$, $y = 022$
Step 3: Find 9's complement of y
Let 9's complement of y is $y' = 888 - y + 1 = 888 - 022 + 1 = 866 + 1 = 867$
Step 4: Add $x + y'$ (which is 9's complement of y)

```
           6 4 5
           8 6 7
(Carry)  1,6 2 3   (Result)
```

Here carry is generated, ignore it, sum after ignoring carry is the final result.
The final result is 623.

Check the answer by direct method: 645 − 022 = 623. This result is same as the result obtained in step 4. Therefore, obtained result is correct.

EXAMPLE 4.35 Perform 45 − 122 subtraction by using 9's complement (number is in nonal form).

Solution:
Step 1: $x = 45$
$y = 122$

Step 2: Here x is of 2-digit number and y is of 3-digit number, so add zero before x to make 3 digits equal to the number of digit in y, $x = 045$.
Now $x = 045$, $y = 122$
Step 3: Find 9's complement of x.
Let 9's complement of y is $y' = 888 - y + 1$
$= 888 - 122 + 1 = 766 + 1 = 767$
Step 4: Add $x + y'$ (which is 9's complement of y)

$$\begin{array}{r} 0\ 4\ 5 \\ 7\ 6\ 7 \\ \hline 8\ 2\ 3 \end{array}$$

Here carry is not generated, so the result is negative. The final result is 823.

Check the answer by direct method: $045 - 122 = -066$. The 9's complement of this number is $888 - 066 + 1 = 822 + 1 = 823$; this result is same as the result obtained in step 4. Therefore, obtained result is correct.

4.7.3 Nonal Multiplication

Nonal multiplication means multiply two nonal numbers in base nine. The multiplication performed in base nine is known as nonal multiplication. In this multiplication carry is taken on 9, like 10 in decimal number.

EXAMPLE 4.36 Perform nonal multiplication $A \times B$ of two given nonal numbers, $A = 334$ and $B = 142$.

Solution: Given $A = 334$ and $B = 142$.
Now perform $A \times B$ as follows:

Carry: On first digit				0 0 0	
On second digit			1 1	1	
On third digit			0 0 0		
A			3	3	4
B			1	4	2
Multiplication by 2 of B to each digit of A			6	6	8
Shift one digit left and multiply 4 of A to each digit of B	1	4	4	7	
Shift two digits left and multiply 1 of A to each digit of B	3	3	4		
Sum of multiplication = Result = Multiplication of $A \times B$	4	8	6	4	8

Multiplication algorithm

Step 1: Let A multiplicand and B multiplier
Step 2: Multiply each digit of A by last digit of B and record it if carry is generated, store it as carry and add it to next digit of A multiplication and do it until the end of digit of A.

Step 3: Shift one digit from left to right and take next digit of B, do as in step 2.
Step 4: Do step 2 to step 3 until you reach to the first digit of B.
Step 5: Perform nonal addition for all multiplication to get the result.

4.7.4 Nonal Division

Nonal division means divide one nonal decimal number by another nonal decimal number in base 9. The division performed in base 9 is known as nonal division. In this division carry is taken on 9, like 10 in decimal number.

EXERCISES

Short Answer Questions

1. What is computer arithmetic?
2. What is binary arithmetic?
3. Why is complement used?
4. Differentiate between decimal and binary arithmetic.
5. What is nonal multiplication?
6. Why is one replaced with zero and zero replaced with one in 1's complement?

Long Answer Questions

1. Perform the following subtraction in 8-bit register by using 1's complement.
   (i) 34 − 56   (ii) 67 − 45   (iii) 60 − 31
   (iv) 31 − 32   (v) 32 − 31
2. Perform the following subtraction in 8-bit register by using 2's complement.
   (i) 35 − 53   (ii) 77 − 54   (iii) 50 − 41
   (iv) 21 − 31   (v) 23 − 13
3. Perform the following subtraction in 8-bit register by using 2's complement.
   (i) 10011 − 1111   (ii) 1011 − 11011   (iii) 11101 − 10101
   (iv) 1001 − 1101   (v) 11111 − 11001
4. Perform the following subtraction in 8-bit register by using 1's complement.
   (i) 1011 − 10111   (ii) 11011 − 101011   (iii) 11001 − 10111
   (iv) 1011 − 11101   (v) 11011 − 11101
5. Perform the following subtraction by using 8's complement.
   (i) 10111 − 110111   (ii) 101011 − 1101011   (iii) 111001 − 101011
   (iv) 11011 − 110101 (v) 111011 − 111101

6. Perform the following subtraction by using 7's complement.
   (i) 10111 − 11111 (ii) 111011 − 1111011 (iii) 1110101 − 1101011
   (iv) 111011 − 1110101 (v) 1111011 − 1111101

7. Perform the following subtraction by using 9's complement.
   (i) 10111 − 11111 (ii) 111011 − 1111011 (iii) 1110101 − 1101011
   (iv) 111011 − 1110101 (v) 1111011 − 1111101

8. Perform the following subtraction by using 9's complement.
   (i) 10111 − 11111 (ii) 111011 − 1111011 (iii) 1110101 − 1101011
   (iv) 111011 − 1110101 (v) 1111011 − 1111101

9. Perform the following subtraction by using 7's complement, the radix of given number is 5.
   (i) 23 − 34 (ii) 21 − 12 (iii) 31 − 13
   (iv) 31 − 32 (v) 24 − 15

10. Perform the following subtraction by using 8's complement, the radix of given number is 5.
    (i) 23 − 34 (ii) 21 − 12 (iii) 31 − 13
    (iv) 31 − 32 (v) 24 − 15

11. Perform the following subtraction using by 9's complement, the radix of given number is 6.
    (i) 25 − 35 (ii) 21 − 15 (iii) 31 − 14
    (iv) 33 − 32 (v) 25 − 25

12. Perform the following binary multiplication of the following numbers. The base of the given number is 8.
    (i) 25 × 35 (ii) 21 × 15 (iii) 31 × 14
    (iv) 33 × 32 (v) 25 × 25

13. Perform the following binary addition of the following numbers. The base of the given number is 8.
    (i) 2 + 35 (ii) 21 + 15 (iii) 31 + 14
    (iv) 33 + 32 (v) 25 + 25

14. Perform the following binary division of the following numbers. The base of the given number is 8.
    (i) 25 ÷ 35 (ii) 21 ÷ 15 (iii) 31 ÷ 14
    (iv) 33 ÷ 32 (v) 25 ÷ 25

15. Perform the following binary division of the following numbers.
    (i) 10111 ÷ 11111 (ii) 111011 ÷ 111011 (iii) 11101 ÷ 11011
    (iv) 111011 ÷ 11101 (v) 1111011 ÷ 11101

16. Evaluate the following expression.
    (i) $(234)_8 + (2ABC8)_{16}$ (ii) $(111011)_2 + (342)_8$ (iii) $(111)_3 + (11011)_2$
    (iv) $(112)_3 + (11101)_3$ (v) $(1111011)_5 + (11101)_5$

17. How many 1's are present in the following expression?
    (i) $(234)_8 \times (23)_8 + (2ABC8)_{16} \times (111011)_2$
    (ii) $1024 \times 512 + 4 \times 8 + 12 \times 24$
18. Perform the following octal addition without converting base.
    (i) $(234)_8 + (23)_8$ (ii) $(1024)_8 + (512)_8 + (47)_8 + (1224)_8$
    (iii) $(564)_8 + (5432)_8$ (iv) $(2213)_8 + (4325)_8$
    (v) $(3456)_8 + (4563)_8$
19. Perform the following hexadecimal addition without converting base.
    (i) $(2A4)_{16} + (23)_{16}$
    (ii) $(1024)_{16} + (5A2)_{16} + (4BA7)_{16} + (1AF4)_{16}$
    (iii) $(5CD4)_{16} + (BCA2)_{16}$
    (iv) $(2EFA)_{16} + (ABC5)_{16}$

Multiple Choice Questions

1. The number of 1's is present in the expression 512×1024 when it is represented in binary number.
   (a) 4 (b) 5
   (c) 2 (d) 6
2. The value of expression $(234)_8 + (23)_8$ is
   (a) 342 (b) 257
   (c) 414 (d) 451
3. Find the value of x if $(234)_8 + x = (23)_8$
   (a) 566 (b) 456
   (c) 576 (d) 457
4. If $(234)_8 + y = (2ABC8)_{16}$ then the value of
   (a) 3AB28 (b) 2AA28
   (c) 11011 (d) 324
5. The 7's complement of $(10101010)_2$ is
   (a) 525 (b) 255
   (c) 552 (d) 155
6. The 2's compliment of $(525)_8$ is
   (a) 010101011 (b) 100110101
   (c) 111010101 (d) 111101011
7. If $x = 1101 \times 11$, the value of x is
   (a) 110110 (b) 100111
   (c) 110101 (d) 110111
8. If $x = (213)_4$ and $y = (323)_4$, then $x + y$ is
   (a) 1202 (b) 1302
   (c) 1201 (d) 1222

9. If $x = (21)_4$ and $y = (32)_4$, then $x \quad y$ is
   (a) 1234
   (b) 1312
   (c) 2312
   (d) 1213
10. If $x = (431)_6$ and $y = (432)_6$, then $x - y$ is
    (a) 454
    (b) 254
    (c) 554
    (d) 354
11. If $x = (4A1)_{16}$ and $y = (4B2)_{16}$, then $x - y$ is
    (a) 45A
    (b) 2B4
    (c) 55F
    (d) FEF

Answers

| 1. (c) | 2. (b) | 3. (a) | 4. (b) | 5. (a) | 6. (a) |
| 7. (a) | 8. (a) | 9. (b) | 10. (c) | 11. (d) | |

BIBLIOGRAPHY

http://www.quadibloc.com/comp/cp02.htm.

Chapter 5

Fundamentals of Boolean Logic and Gates

5.1 INTRODUCTION

Boolean logic is defined as a type of decision making used by computers to decide whether a statement is true or false. The four main Boolean operators are used to evaluate whether a statement is true or false. These operators are given below:

(i) And operator (&&): The result is true if both operands are true.
(ii) OR operator (||): The result is true if at least one operand is true.
(iii) XOR: The result is true if ONLY one operand is true.
(iv) Negation operator (!): The result is true if a single operand is false.

Let us consider A is true, B is true and C is false, then **A && B**: As per definition if both operands are true then the result is true. In this A and B both are true, therefore this expression is true.

$A \mid\mid C$: In this expression A is true or B is true, then the result is true. In this example A and B both are true, therefore the expression is again true.

A XOR B: In this expression A or B but not both, to be true. In this example both are true therefore the expression is false.

!A: If A is true then the result will be false and if A is false then the result is true. In this example A is true so the result is false.

!C: If C is false the result is true and C is true then the result is false. In this example C is false therefore the result is true.

5.1.1 Power Set

Power set is defined as set of all the subsets of a set. Consider a set $X = \{x_1, x_2, x_3, x_4\}$, the all subset of this set are $\{x_1\}$, $\{x_2\}$, $\{x_3\}$, $\{x_4\}$, $\{x_1, x_2\}$, $\{x_1, x_3\}$, $\{x_1, x_4\}$, $\{x_2, x_3\}$, $\{x_2, x_4\}$, $\{x_3, x_4\}$ and = $\{x_1, x_2, x_3, x_4\}$ is also subset of = $\{x_1, x_2, x_3, x_4\}$ but is not a proper subset and an empty set { } is also a subset of X set. If you are listing all subset of X set, that all sets will be the power set of X set. The power set is denoted by $P(X)$.

The power set $P(X) = \{\{\ \}, \{x_1\}, \{x_2\}, \{x_3\}, \{x_4\}, \{x_1, x_2\}, \{x_1, x_3\}, \{x_1, x_4\}, \{x_2, x_3\}, \{x_2, x_4\}, \{x_3, x_4\}, \{x_1, x_2, x_3, x_4\}\}$

You can use various ways to select the elements of set, the order of the element does not matter while determining the power set, including all or none. In general term, if the set has n number of elements then the power set will have 2^n elements.

EXAMPLE 5.1 If the set X is defined as $X = \{x_1, x_2, x_3, x_4\}$, find the total elements of power set.

Solution: The elements of X are 4 (four), so the total elements of power set are $2^4 = 16$.

Notation of power set

The number of elements of a set defined as $|X|$, where X is the name of set. The power set is defined as

$$|P(X)| = 2^n$$

EXAMPLE 5.2: Y is defined as $Y = \{1, 2, 3\}$. What are total elements of power set?

Solution: Here, Y having 3 elements, therefore total number of elements in the power set are $|P(Y)| = 2^3 = 8$.

5.2 EXAMPLES

5.2.1 Boolean Algebra

Boolean algebra deals with two or more variables which value is true or false (0 or 1) and functions that find its use in digital computers, the two states true or false is used called binary systems. George Boole is a mathematician who gave the concept of Boolean algebra. He used 1 for true, 0 for false, . (Dot) for AND and + (Plus) for OR.

Consider the statement: "I will appear in a test If I have a pen or I have a pencil". This statement shows the fact that the proposition "appear in a test" depends on two other propositions "have a pen" and "have a pencil". Any one of these two propositions can be either true or false hence the table of all possible situations is given below.

The two operators are defined as OR operator represented by + (Plus) and AND operator represented by . (Dot). The NOT operator is the third operator in Boolean algebra which inverts the input. Table 5.1 for above situation is given below where Not A is represented as A' and Not B is represented as B'

Table 5.1 OR operator truth table

Pen	Pencil	Appear in Test
No	No	No
Yes	No	Yes
No	Yes	Yes
Yes	Yes	Yes

Let A = pen, B = pencil, Yes = 1, No = 0 and X = appear in test. Table 5.1 can be represented in binary system as below:

Table 5.2 OR operator truth table in binary system

A	B	X
0	0	0
1	0	1
0	1	1
1	1	1

The three operators are the basic operators used in Boolean algebra and from which more complicated Boolean expressions may be written. The Boolean expression from Table 5.2 is written as $X = X = A.B' + A'.B + A.B$

5.3 POSTULATE

The postulate is defined as something taken as self-evident or assumed without proof as a basis for reasoning. In Mathematics, Logic, a proposition that requires no proof, being self-evident, or that is for a specific purpose assumed true, and the postulate is generally used in the proof of other propositions; axiom. Postulate is universally accepted. For example, $4 \times 3 = 3 \times 4$ is universally accepted. This is also known as an axiom. There is no need to prove this fact. Assume A, B and C are logical states that can have the values 0 (false) and 1 (true). "+" means OR, "·" means AND and NOT [A] means NOT A.

The list of postulates are given below:

(i) Identity postulate: $A + 0 = A$, for all $A \in X$
(ii) Complement postulate: $A + A' = 1$ and $A.A' = 0$
(iii) Commutative law: $A + B = B + A$ and $A.B = B.A$
(iv) Associative law: $A + (B + C) = (A + B) + C$
(v) Distributive law: $A + (B.C) = (A + B).(A + C)$ and $A.(B + C) = (A.B) + (A.C)$

5.4 THEOREM

According to Mathematics, the theorem may be defined as a proposition that has been or is to be proved on the basis of explicit assumptions. The list of theorem of Boolean algebra is given as follows:

(i) $A + 0 = A$ where $A = \{0, 1\}$
(ii) $A.1 = A$ where $A = \{0, 1\}$
(iii) $A + 1 = 1$ where $A = \{0, 1\}$
(iv) $A.0 = 0$ where $A = \{0, 1\}$
(v) $A + A = A$ where $A = \{0, 1\}$
(vi) $A.A = A$ where $A = \{0, 1\}$
(vii) $(A')' = A$ where $A = \{0, 1\}$
(viii) $A + A' = 1$ where $A = \{0, 1\}$
(ix) $A.A' = 0$ where $A = \{0, 1\}$
(x) $A + AB = A$
(xi) $A(A + B) = A$
(xii) $(A + B') B = AB$
(xiii) $AB' + B = A + B$
(xiv) $(A + B + C + D...)' = A'.B'.C'.D'...$ (De Morgan's law defined on + operator)
(xv) $(A.B.C.D....)' = A' + B' + C' + D'+...$ (De Morgan's law defined on . operator)

In general form the De Morgan's law defined as follows:
If there are n variables like $A_1, A_2, A_3,...,A_n$, then the De Morgan's law can be defined as follows:
$\{F(A_1, A_2, A_3, ..., A_n, \text{false, true, or, and})\}' = F(A_1', A_2', A_3', ..., A_n', \text{true, false, and, or})$, where true = 1, false = 0, and = ., or = +.

(xvi) $(A + B).(A' + C) = A.C + A'.B$
(xvii) $AB + A'C = (A + C) (A' + B)$

5.5 BOOLEAN FUNCTION

The Boolean function is defined as a function into the form of $F:A^n \rightarrow A$, where $A = \{\text{false, true}\}$ is defined as Boolean domain and n is a positive integer. In the case where $n = 0$, then F is called the constant function, and its value became constant element of A. The Boolean domain has been seen in the field of logic as $A = \{1, 0\}$, where 1 stands for true and 0 for false. It is found from count of Boolean functions, there are $2^{(2^n)}$ Boolean functions on n variables.

EXAMPLE 5.3 How many Boolean functions from 4 variables for the count of Boolean function.

Solution: We know the number of Boolean function = $2^{(2^n)}$ where n is a variable.
Here, the value of n is 4 so the number of Boolean function is

$2^{(2^4)} = 2^{2*2*2*2} = 2^{16} = 2^{10} * 2^6 = 1024 * 64 = 65536.$

The Boolean function is defined in another way by introducing the dependent variable and independent variable. Let a and b are independent variables and y is a dependent variable, the relationship between independent variable and dependent variable is given as follows:

$y = \sum(a, b)$ or $y = \prod(a, b)$, if the values of a and b are known then we can find the value of y. This relationship can be represented mathematically as $y = f(a, b)$. This function may be in sum of product or product of sum. This function is represented in sum of product, if $a = \{0, 1\}$ and $b = \{0, 1\}$ then find the cross product as follows:

$a \times b = \{(0, 0), (0, 1), (1, 0), (1, 1)\}$, therefore, the function in product of sum can be shown as $f(a, b) = \sum \{(0, 0), (0, 1), (1, 0), (1, 1)\} = 00 + 01 + 10 + 11$, in this equation the first digit represents a and the second digit represents b. Replace the first digit with a and the second digit with b and 0 with normal variable and 0 with complement variable as follows:

$f(a, b) = a'.b' + a'.b + a.b' + a.b$, the same function can be represented in product of sum as follows:

$f(a, b) = \prod \{(0, 0), (0, 1), (1, 0), (1, 1)\} = (0 + 0).(0 + 1).(1, 0).(1, 1)$

$f(a, b) = (a + b).(a + b').(a'.b).(a' + b')$

EXAMPLE 5.4 Write the truth table of the function $f(a, b) = a'.b' + a'.b + a.b' + a.b$.

Solution:

a	b	$f(a, b) = a'.b' + a'.b + a.b' + a.b$
0	0	0
0	1	1
1	0	1
1	1	1

EXAMPLE 5.5 Find the Boolean function for $F1$ and $F2$ from the given truth table

x	y	z	F1	F2
0	0	0	1	0
0	0	1	0	1
1	0	0	1	1
1	0	1	0	1
1	1	0	1	0
1	1	1	1	1

Solution: Scan $F1$ from top to bottom, if 1 is encountered then write the corresponding variables in the term and sum all the terms to get the function in sum of product.

$$F1(x, y, z) = x'y'z' + xy'z' + xyz' + xyz$$

Similarly

$$F2(x, y, z) = x'y'z + xy'z' + xy'z + xyz$$

These functions are also written in product of sum as follows:

$$F1(x, y, z) = (x + y + z)(x' + y + z)(x' + y' + z)(x' + y' + z')$$
$$F2(x, y, z) = (x + y + z)(x' + y + z)(x' + y + z')(x' + y' + z')$$

5.6 LOGIC GATES

According to webopedia, the logic gate is a type of circuit. In another way you can say that the collection of transistors and resistors that regulates the flow of electricity that determines the Boolean logic computer use to make complex logical decisions. The components of logic gates are transistor and register. There are three types of logic gates, namely, AND, OR and NOT.

5.6.1 AND Logic Gate

The AND gate is used to find the product of two or more variables. Let a, b and c are three variables which are written as $a.b.c$ in AND gate. The logical circuit of AND gate is given in Figure 5.1.

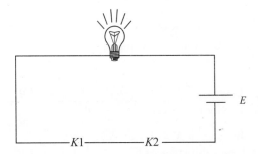

Figure 5.1 AND gate logical circuit diagram.

The truth table of AND gate is given in Table 5.3 and its symbol is given in Figure 5.2.

Table 5.3 AND gate truth table

K1	K2	X = Status of the Bulb, 1 = On, 0 = Off
0	0	0
0	1	0
1	0	0
1	1	1

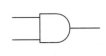

Figure 5.2 AND gate symbol.

The state equation of AND gate is $X = K1.K2$, because bulb is ON only when both keys $K1$ and $K2$ are in the state of ON.

5.6.2 OR Logic Gate

The OR gate is used to find the sum of two or more variables. Let a, b and c are three variables which are written as $a + b + c$ in OR gate. The logical circuit of OR gate is given in Figure 5.3.

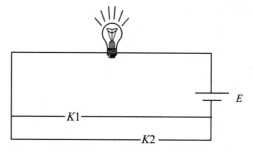

Figure 5.3 OR gate logical circuit diagram.

The truth table of OR gate is given in Table 5.4 and its symbol is given in Figure 5.4.

Table 5.4 OR gate truth table

K1	K2	X = Status of the Bulb, 1 = On, 0 = Off
0	0	0
0	1	1
1	0	1
1	1	1

Figure 5.4 OR gate symbol.

The state equation of OR gate is $X = K1 + K2$, because the bulb is ON only when any one from keys $K1$ and $K2$ are in the state of ON.

$$X = K1'K2 + K1(K2' + K1K2)$$
$$= K1'K2 + K1(K2' + K2) = K1'K2 + K1.1 \text{ where } K2' + K2 = 1$$
$$= K1'K2 + K1 \text{ where } K1.1 = K1$$
$$= K1 + K2 \text{ where } K1'K2 + K1 = K1 + K2$$

5.6.3 NOT Logic Gate

The NOT gate is used to find the compliments of a given input, means if input is 1 the output will be 0 and if input is 0 then output will be 1. Let a be the input and the output will be a'. The logical circuit of NOT gate is given in Figure 5.5.

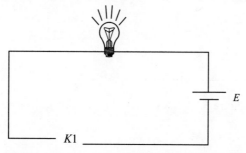

Figure 5.5 NOT gate logical circuit diagram.

The truth table of NOT gate is given in Table 5.5 and its symbol is given in Figure 5.6.

Table 5.5 NOT gate truth table

K1	X = Status of the Bulb, 1 = On, 0 = Off
0	1
1	0

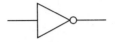

Figure 5.6 NOT gate symbol.

The state equation of NOT gate is $X = K1'$, because the bulb is ON only when $K1$ is 0 and bulb is OFF when $K1$ is 1.

5.6.4 NAND Gate

The NAND gate is the combination of AND gate and NOT gate. The logical circuit diagram of NAND gate is given in Figure 5.7. In this diagram the logical diagram of AND gate and logical diagram of NOT gate are added to each other. The output of logical circuit will be the input for logical circuit of NOT gate. The output of NOT gate will be the output of NAND gate.

$$AND + NOT = NAND$$

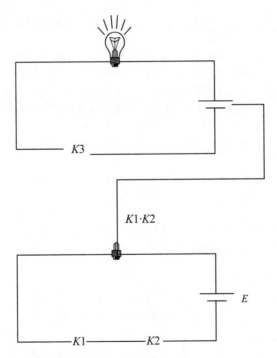

Figure 5.7 NAND gate logical circuit diagram.

Fundamentals of Boolean Logic and Gates

The truth table of NAND gate is given in Table 5.6 and its symbol is given in Figure 5.8.

Table 5.6 NAND gate truth table

K1	K2	K3 = K1.K2	X = Status of the Bulb, 1 = On, 0 = Off K3'
0	0	0	1
0	1	0	1
1	0	0	1
1	1	1	0

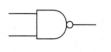

Figure 5.8 NAND gate symbol.

The equation of NAND gate is $X = (K1.K2)'$

5.6.5 NOR Gate

The NOR gate is the combination of OR gate and NOT gate. The logical circuit diagram of NOR gate is given in Figure 5.9. In this diagram the logical diagram of OR gate and logical diagram of NOT gate are added to each other. The output of logical circuit will be the input for logical circuit of NOT gate. The output of NOT gate will be the output of NOR gate.

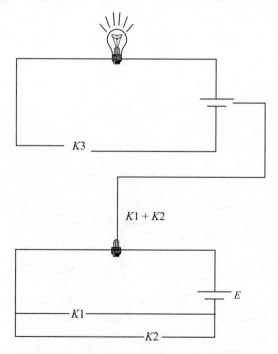

Figure 5.9 NOR gate logical circuit diagram.

The truth table of NOR gate is given in Table 5.7 and its symbol is given in Figure 5.10

Table 5.7 NOR gate truth table

K1	K2	K3 = K1 + K2	X = Status of the Bulb, 1 = On, 0 = Off K3'
0	0	0	1
0	1	1	0
1	0	1	0
1	1	1	0

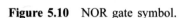

Figure 5.10 NOR gate symbol.

The equation of NAND gate is $X = (K1 + K2)'$

5.6.6 Exclusive-OR Gate

The exclusive-OR gate produces output one or true when any one output is zero (false) and another input one (true). If both inputs are same then the output will be false, means output will be zero. Let two inputs are A and B and output is X, the truth table is given in Table 5.8. This gate is also known as XOR gate. The XOR or exclusive-OR gate symbol is given in Figure 5.11.

Table 5.8 XOR gate truth table

A	B	X
0	0	0
0	1	1
1	0	1
1	1	0

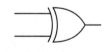

Figure 5.11 XOR gate symbol.

The equation of exclusive-OR or XOR is given below.

$$X = A \oplus B$$

5.6.7 Exclusive-NOR Gate

The exclusive-NOR gate is a combination of exclusive-OR and NOT gates. The output of exclusive-NOR gate will be true or one, if both inputs will be either zero or one. The truth table of exclusive-NOR gate is given in Table 5.9. The symbol of exclusive-NOR gate is given in Figure 5.12.

Table 5.9 Exclusive-NOR gate truth table

A	B	$X = A \oplus B$	$X' = A \odot B$
0	0	0	1
0	1	1	0
1	0	1	0
1	1	0	1

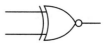

Figure 5.12 Exclusive-NOR gate symbol.

The equation of exclusive-NOR gate is given below.

$$X = A \odot B$$

5.7 CIRCUIT OF BOOLEAN FUNCTION

The Boolean function is represented in the form circuit by Dot (.) operator with AND gate, Plus (+) operator with OR gate and negation with NOT gate. These are four fundamental gates used to convert the Boolean function into logical circuit.

EXAMPLE 5.6 Draw the logic circuit for the Boolean function $f(x, y, z) = xy + x'yz + y'z$.

Solution: The xy is representing by AND is as follows:

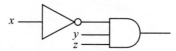

The $x'yz$ represented by AND gate is as follows:

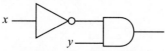

The $y'z$ represented by AND gate is as follows:

To get the final result these all gates are added with the one OR gate as shown below.

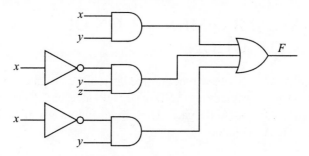

EXAMPLE 5.7 Draw logical circuit of the Boolean function $f(x, y, z) = \sum(2, 3, 4)$.

Solution: We can write the given Boolean function as follows:

$$f(x, y, z) = 010 + 011 + 100$$
$$f(x, y, z) = x'yz' + x'yz + xy'z$$

This function can be represented in logic circuit as follows:

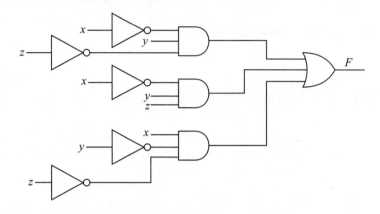

EXAMPLE 5.8 Draw the Boolean function $f(x, y, z) = \prod(2, 3, 4)$.

Solution: The given function can be represented as follows:

$$f(x, y, z) = (0 + 0 + 1)(0 + 1 + 1)(1 + 0 + 0)$$
$$f(x, y, z) = (x + y + z')(x + y' + z')(x' + y + z)$$

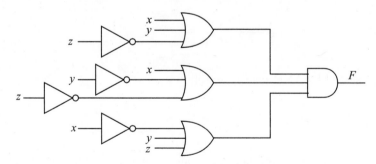

5.8 NAND GATE IMPLEMENTATION OF BOOLEAN FUNCTION

The NAND implementation of Boolean function means replace all gates in circuit with NAND gates only. Use NAND gate in place of AND gate, OR gate and NOT gate. The following NAND gate circuit equivalent to AND gate. The circuit output given in Figure 5.13 will be same as the output of the AND gate.

Fundamentals of Boolean Logic and Gates **103**

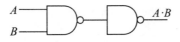

Figure 5.13

The following NAND gate circuit replaces the OR gate. The output of this circuit is same as the output of OR gate. The NAND gate equivalent to OR gate is given in Figure 5.14.

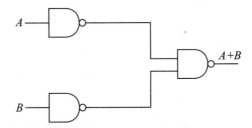

Figure 5.14 NAND gate equivalent to OR gate.

The following NAND gate circuit replaces the NOT gate. The output of this circuit is same as the output of NOT gate. The NAND gate equivalent to NOT gate is given in Figure 5.15.

Figure 5.15 NAND gate equivalent to NOT gate.

5.8.1 Procedure for NAND Gate Implementation

The following steps are the procedure to implement the Boolean circuit with NAND gates.

Step 1: Replace all gates in circuit with NAND gate equivalent to AND gate, OR gate and NOT gate. Make the dotted line square and rectangle of replace gate. This is known as level first.

Step 2: Replace the dotted square and rectangle gates with one NAND gate. This is level 2.

Step 3: If the output of the NAND gate implemented circuit is same as the output of original circuit, then the NAND implemented circuit is the final NAND implemented circuit. Else go to Step 4.

Step 4: Add or remove NAND gate to the implemented circuit in Step 3 to get the same output as the original circuit output. This is level 3.

EXAMPLE 5.9 Implement the following function with NAND gate.

$$f(x, y, z) = xy + y'z' + yz$$

Solution: The circuit diagram of the Boolean function is given below:

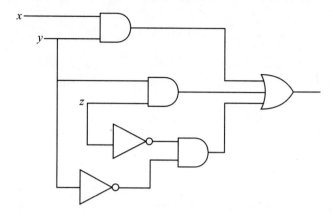

Now replace each gate in the above diagram with the NAND equivalent gate of AND, OR and NOT.

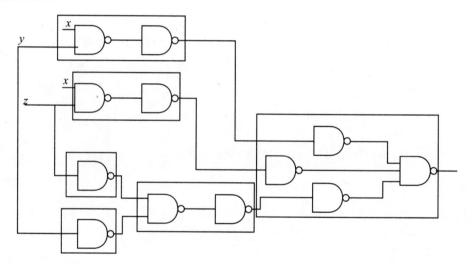

Now replace each square and rectangle with one NAND gate as follows:

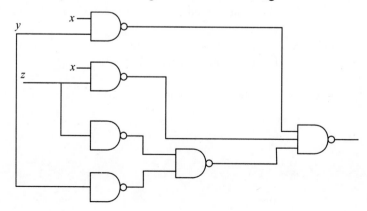

The output of the above circuit is $[(x.y)' + (x.z)' + y'.z']'$. This result is not the same as the output of the original circuit result. The original circuit output is $xy + y'z' + yz$. Therefore, add some NAND gates in the above circuit to get the desired result as follows:

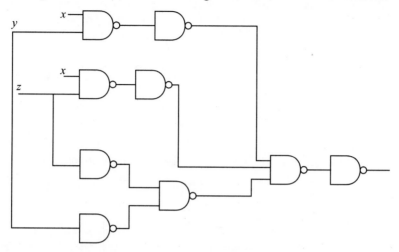

EXAMPLE 5.10 Implement the following circuit with NAND gate.

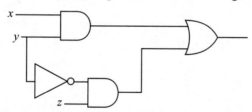

Solution: Replace all gates in the given diagram with NAND gate equivalent of AND, OR and NOT gates in the given circuit. Replace all gates in the given diagram with NAND gate equivalent of each gate.

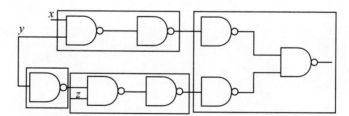

Now replace each square and rectangle with one NAND gate in the above diagram.

The output of the above circuit is same as the output of the given circuit. Therefore, the above circuit is NAND implemented circuit.

EXAMPLE 5.11 Implement the following circuit with NAND gate.

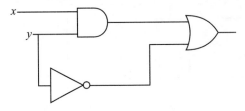

Solution: The output of the given circuit is $xy + y'$. Replace all gates in the given diagram with NAND gate equivalent of AND, OR and NOT gates in the given circuit.

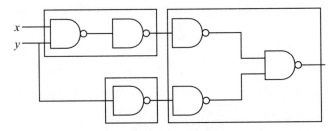

Replace each square and rectangle with one NAND gate in the above diagram.

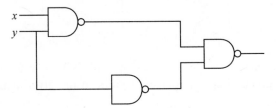

The output of the above diagram is $[(xy)'(y)']' = xy + y$. This result is not same as the original result. Therefore, some NAND gates will be added in the above circuit to get the original result. The following circuit is NAND implemented of the given circuit.

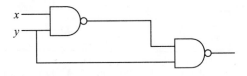

5.9 NOR GATE IMPLEMENTATION OF BOOLEAN FUNCTION

The NOR implementation of Boolean function means replace all gates in circuit with NOR gates only. Use NOR gate in place of AND gate, OR gate and NOT gate. The following NOR gate circuit is equivalent to AND gate. The circuit output given below will be same as the output of the AND gate. The NOR gate equivalent to AND gate is given in Figure 5.16.

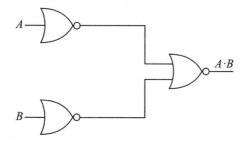

Figure 5.16 NOR gate equivalent to AND gate.

The NOR gate equivalent to OR gate is given in Figure 5.17.

Figure 5.17 NOR gate equivalent to OR gate.

The NOR gate equivalent to NOT gate is given in Figure 5.18.

Figure 5.18 NOR gate equivalent to NOT gate.

5.9.1 Procedure for NOR Gate Implementation

The following steps are procedure to implement the Boolean circuit with NOR gates.

Step 1: Replace all gates in circuit with NOR gate equivalent to AND gate, OR gate and NOT gate. Make the dotted line square and rectangle of replace gate. This is known as level 1.

Step 2: Replace the dotted square and rectangle gates with one NOR gate. This is level 2.

Step 3: If the output of the NOR gate implemented circuit is same as the output of original circuit, then the NOR implemented circuit is the final NOR implemented circuit. Else go to Step 4

Step 4: Add or remove NOR gate to the implemented circuit in the Step 3 to get the same output as the original circuit output. This is level 3.

EXAMPLE 5.12 Implement the following function with NOR gate.

$$f(x, y, z) = xy + y'z'$$

Solution: The circuit diagram of the Boolean function is given below:

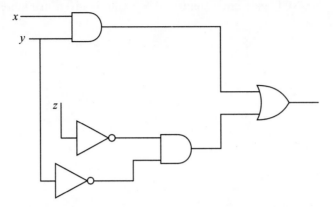

Now replace each gate in the above diagram with the NOR equivalent gate of AND, OR and NOT.

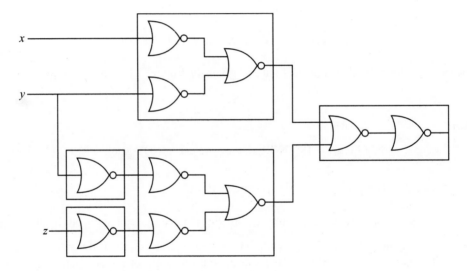

Now neither replaces each square and rectangle with one NOR gate in the above diagram.

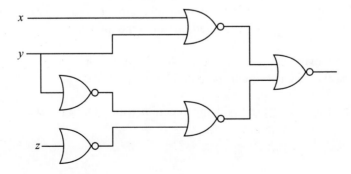

The output of the above diagram is $[(x + y)' + (y' + z')']' = x'y' + yz$. This result is not same as the original result. Therefore, some NOR gate will be added in the above circuit to get the original result. The following circuit is NOR implemented of the given Boolean function.

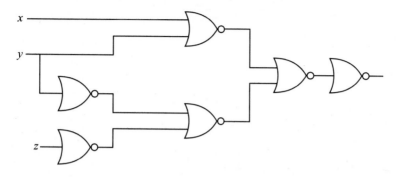

EXAMPLE 5.13 Implement the following circuit with NOR gate.

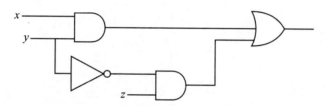

Solution: The output of the given circuit is $xy + y'z$. Replace each gate in the circuit given in the example with the NOR equivalent gate of AND, OR and NOT. You get as follows:

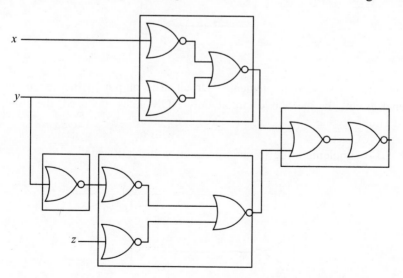

110 Digital Logic Design

Now replace each square and rectangle with one NOR gate only to get the diagram given below.

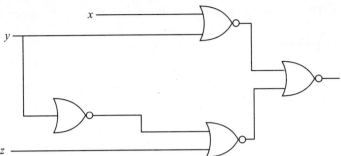

Now the output of the above circuit is $[(x + y)' + (y' + z)']'$. This output is not the same as the original circuit output. Therefore, add and remove some NOR gates in the above circuit to get the output same as the original output.

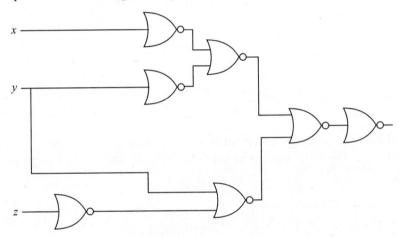

The above circuit is NOR implemented circuit. The output of the above circuit is same as the original circuit. In the above circuit only NOR gates are used, therefore, it is NOR implemented circuit.

EXAMPLE 5.14 Implement the following circuit with NOR gate.

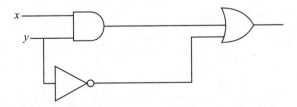

Solution: The Boolean function from the above circuit is $f(x, y) = xy + y'$. Replace each gate in the circuit given in the example with the NOR equivalent gate of AND, OR and NOT. You get as follows:

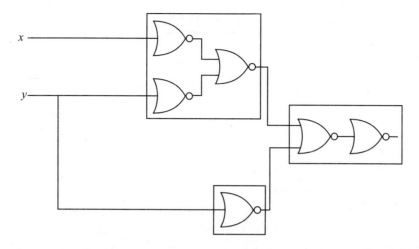

Now replace each square and rectangle in the above diagram with one NOR gate only to get the diagram given below.

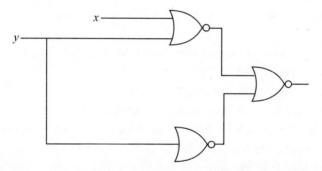

Now the output of the above circuit is $[(x + y)' + y']'$. This output is not the same as original circuit output. Therefore, add some NOR gate in the above circuit to get the output same as original output.

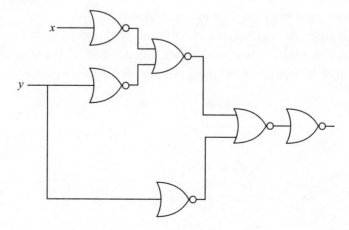

EXERCISES

Short Answer Questions

1. What do you mean by Boolean logic?
2. What is Boolean function?
3. What are fundamental gates?
4. List the universal gates.
5. What are elements of logic gate?
6. Differentiate between postulate and theorem.
7. What is exclusive-OR gate?
8. Draw truth table of exclusive-NOR gate.
9. Define Boolean algebra.
10. Why is Boolean algebra used?

Long Answer Questions

1. Why is Boolean logic played an important role in circuit design?
2. What are fundamental gates? Explain each gate.
3. Prove $x + x' = x + y$ without using truth table.
4. Prove $x + x = x$ and $x.x = x$ without using truth table.
5. How many Boolean functions from 6 variables for the count of Boolean function?
6. State and prove the De Morgan's law.
7. Write the truth table of the function $f(a, b) = x'.y'.z + x'.y.z + x.y' + x.y.z$
8. Draw the logic circuit for the Boolean function $f(x, y, z) = xyz + x'y + xy'z$.
9. Draw the logic circuit of the following Boolean function:
   $F1(x, y, z) = (x + y + z)(x' + y)(x' + y')(x' + y' + z')$
10. Why is Boolean algebra required in circuit design?
11. Draw truth table and logic circuit of AND and NOT gates.
12. Draw truth table, logic circuit of OR gate and also find its Boolean function.
13. Find the Boolean function from the given logical diagram below.

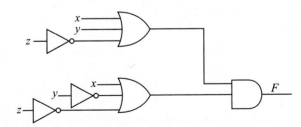

14. Draw logical circuit of the Boolean function $f(x, y, z, w) = \Sigma(2, 3, 4, 7, 9, 13)$.
15. Find the Boolean function from the given logical circuit below:

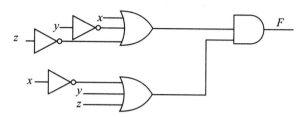

16. Draw logical circuit of the Boolean function $f(x, y, z, w) = \Sigma(2, 3, 4, 7, 9, 13)$.
17. Find the Boolean function from the given logical circuit below:

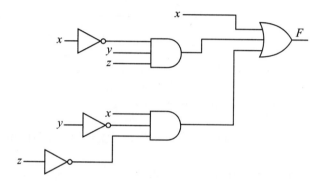

18. Draw the logical circuit from the following truth table:

x	y	z	F1	F2
0	0	0	1	1
0	0	1	1	0
1	0	0	1	1
1	0	1	0	0
1	1	0	0	0
1	1	1	1	1

19. Find Boolean function of $F1$ and $F2$ from the table given in Problem 18.
20. Prove $x \cdot 1 = x$ and $x + 1 = 1$.
21. Find the truth table and Boolean function of the following circuit.

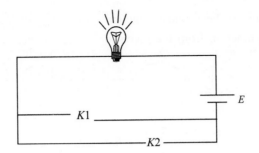

22. Name the logical gate represented by the following circuit and specify why.

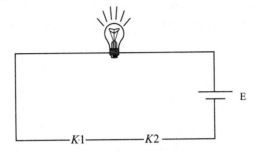

23. Draw the circuit for the Boolean function $f(x, y, z) = \Sigma(2, 3, 4)$ and also implement it with NAND gate.

24. Implement the following circuit with NAND gate.

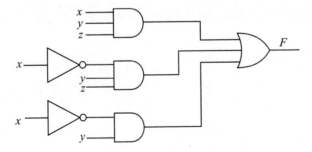

25. The circuit of Boolean function is given below. Implement the following circuit diagram with NOR gate.

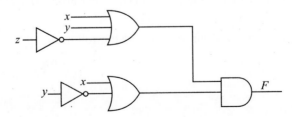

Fundamentals of Boolean Logic and Gates   115

26. Draw the circuit for the Boolean function $f(x, y, z) = \sum(1, 2, 5, 4)$ and also implement it with NOR gate.
27. Find the new circuit using AND, OR and NOT gates from the given circuit whose output is neither same as the given circuit without using NOR gate.

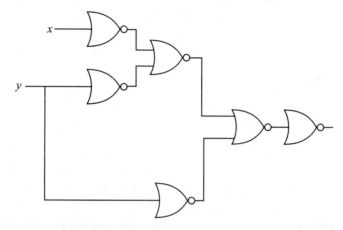

28. Draw the circuit diagram for the Boolean function $f = xy + yz + xy'z$ using
    (i) AND, OR and NOT gates (ii) OR and NOT gates
    (iii) AND and NOT gates
29. What is the NAND equivalent of the following gates?
    (i) AND (ii) OR
    (iii) NOT
30. What is the NAND equivalent of the following gates?
    (i) OR gate (ii) NOT gate
    (iii) AND
31. Covert the following circuit using AND, OR and NOT gates without effecting the output of this circuit.

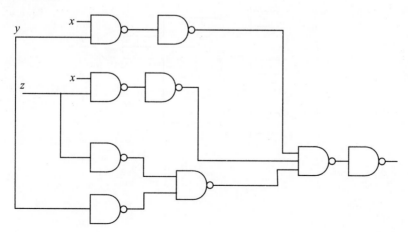

32. Implement the following Boolean function with NAND and NOR gates.

   (i) $f(x, y, z) = \prod(1, 2, 5, 6, 7)$

   (ii) $f(x, y, z) = \prod(0, 2, 4, 5, 6, 7)$

   (iii) $f(x, y, z) = (y + z)(x + y)(x' + y)$

   (iv) $f(x, y, z) = (y' + z)(x + y)(x + y')$

   (v) $f(x, y, x) = x'y + yz + xy' + x'$

   (vi) $f(x, y, z) = y'z + xz + xy' + xy$

Multiple Choice Questions

1. Which of the following is universal gate?
   - (a) AND
   - (b) OR
   - (c) NOT
   - (d) NAND

2. The NAND gate is a combination of
   - (a) AND and NOT
   - (b) OR and NOT
   - (c) AND and OR
   - (d) AND, OR and NOT

3. Which of the following is De Morgan's law?
   - (a) $(x.y)' = x' + y'$
   - (b) $(x + y)' = x.y$
   - (c) (a) and (b)
   - (d) $x.y = x + y$

4. Which of the following is not true?
   - (a) $x + x' = 1$
   - (b) $x + x'y = x + y$
   - (c) $x + x = 2x$
   - (d) $x.x' = 0$

5. Which of the following is true?
   - (a) $x + x' = 0$
   - (b) $x + x'y = x + y$
   - (c) $x + x = 2x$
   - (d) $x.x' = 1$

6. The output of the following diagram is

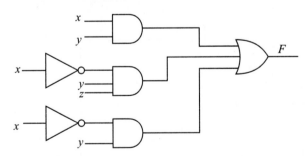

   - (a) $y + x'yz$
   - (b) $x + x'y$
   - (c) $xy + xyz$
   - (d) $x'y + xy$

7. If $A + 1 = x$, then what is the value of x?
   (a) 0
   (b) 1
   (c) A
   (d) (b) and (c)
8. How many logic gates are required to draw the logical circuit of the Boolean function $f(x, y, z) = x'yz' + x'yz + xy'z'$
   (a) 6
   (b) 5
   (c) 7
   (d) 8
9. Which of the following is equivalent to $xy + x'y + x'yz$
   (a) $y + x'yz$
   (b) $x.y + y'$
   (c) $x + yx$
   (d) $x.y + y$
10. Which of the following gate is represented by this diagram?

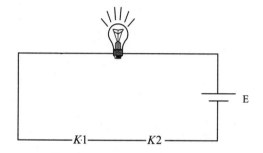

   (a) AND
   (b) OR
   (c) NOT
   (d) NAND

11. Which of the following gate is represented by this diagram?

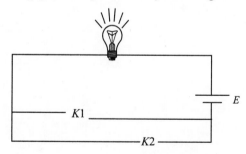

   (a) AND
   (b) OR
   (c) NOT
   (d) NAND

Answers

1. (d) 2. (a) 3. (c) 4. (c) 5. (b) 6. (a)
7. (b) 8. (c) 9. (a) 10. (a) 11. (b)

BIBLIOGRAPHY

Angela Bradley, *Boolean Logic*, About.com Guide.

Givant, S. and Halmos, P., *Introduction to Boolean Algebras (Undergraduate Texts in Mathematics)*, Springer, ISBN-13: 978-0387402932.

Mendelson, E., *Introduction to Boolean Algebra and Switching Circuits*. McGraw-Hill, New York, 1973.

CHAPTER 6

Simplification of Boolean Function

6.1 INTRODUCTION

The simplification of Boolean function is done for reducing the size of circuit. The simplification is also helpful in reducing the cost of logical circuit.

6.2 SIMPLIFICATION BY BOOLEAN ALGEBRA

The postulate and theorems are used to simplify the Boolean function. The Boolean algebra method is more suitable for simplification of minimum variable Boolean function. The Boolean function having more variables, then the Boolean algebra method will not be good. But you can simplify any complex Boolean function by this method.

EXAMPLE 6.1 Simplify the Boolean function $f(x, y, z) = x'yz + xy + xy'z + xyz'$.

Solution:
$$f(x, y, z) = x'yz + xyz + xy'z' + xyz'$$
$$= yz(x' + x) + xz'(y' + y)$$
$$= yz.1 + xz'.1 \text{ from } x' + x = 1$$
$$= yz + xz' \text{ from } x.1 = x$$

The simplified function is $f(x, y, z) = yz + xz'$

EXAMPLE 6.2 Simplify the Boolean function
$$f(x, y, z) = (x' + y + z)(x + y + z)(x + y' + z)(x + y + z')$$

Solution: $f(x, y, z) = (x' + y + z)(x + y + z)(x + y' + z)(x + y + z')$

Taking compliment both sides of the given Boolean function as follow:
$$f'(x, y, z) = [(x' + y + z)(x + y + z)(x + y' + z)(x + y + z')]'$$

Using De Morgan's law
$$f'(x, y, z) = (x' + y + z)' + (x + y + z)' + (x + y' + z)' + (x + y + z')'$$
$$= x.y'.z' + x'.y'.z' + x'.y.z' + x'.y'.z$$

$$= xz'(y' + y) + x'y'(z' + z)$$
$$= xz'.1 + x'y'.1 \text{ by } x + x' = 1$$
$$= xz'.1 + x'y'.1 \text{ by } x.1 = x$$
$$f'(x, y, z) = xz' + x'y'$$

Now taking compliment both sides again to get the simplified Boolean function
$$[f'(x, y, z)] = (xz' + x'y')'$$
$$f(x, y, z) = (xz')'(x'y)'$$
$$f(x, y, z) = (x' + z)(x + y')$$

The simplified function is $f(x, y, z) = (x' + z)(x + y')$.

6.3 CANONICAL FORMS OF BOOLEAN ALGEBRA

The function is in simplified form, and then the canonical forms are used to bring the simplified Boolean function into its original Boolean function. In the canonical form, introduce the eliminated variable from the given function by using postulate and theorem of Boolean algebra.

EXAMPLE 6.3 Find canonical forms of the Boolean function $f(x, y, z) = yz + xz'$.

Solution: In the given Boolean function x is missing in first term and y in next term. Introduce x in first term and y in next term to convert the given Boolean function into canonical form. Multiply both terms by 1.

$$f(x, y, z) = yz.1 + xz'.1$$

We know $x + x' = 1$, the x term is missing in first term of Boolean function so put $x + x'$ in place of 1 in first term and y is missing from second term so replace 1 by $y + y'$ of the Boolean function.

$$f(x, y, z) = yz(x + x') + xz'(y + y')$$
$$= xyz + x'yz + xyz' + xy'z'$$

$f(x, y, z) = xyz + x'yz + xyz' + xy'z'$ this function is now in canonical form.

EXAMPLE 6.4 Find canonical forms of the Boolean function $f(x, y, z) = yz + xz'$ using truth table.

Solution:

x	y	z	$f(x, y, z) = yz + xz'$
0	0	0	0
0	0	1	0
0	1	0	0
0	1	1	1
1	0	0	1
1	0	1	0
1	1	0	1
1	1	1	1

In the above table scan the function column from top to bottom, when 1 is encountered, write the corresponding term and sum all the terms to get the canonical form.

$f(x, y, z) = x'yz + xy'z' + xyz' + xyz$, this is the canonical form of the above Boolean function.

EXAMPLE 6.5 Find canonical form of the Boolean function $f(x, y, z) = (y' + z')(x' + z)$.

Solution: $f(x, y, z) = (y' + z')(x' + z)$, in this Boolean function x is missing from the first term and y is missing from the second term. We have to introduce the x in the first term and y in the second term to get the Boolean function in canonical form.

$$f(x, y, z) = (y' + z' + 0)(x' + z + 0')$$
$$= (y' + z' + x.x')(x' + z + y.y') \text{ by } x.x' = 0 \text{ and } y.y' = 0$$
$$= (y' + z' + x.x')(x' + z + y.y')$$
$$= (y' + z' + x)(y' + z' + x')(x' + z + y)(x' + z + y') \text{ by distributive law}$$
$$= (x + y' + z')(x' + y' + z')(x' + y + z)(x' + y' + z)$$

Now the Boolean function $f(x, y, z) = (x + y' + z') (x' + y' + z') (x' + y + z) (x' + y' + z)$ is in the canonical form.

EXAMPLE 6.6 Find canonical form of the Boolean function $f(x, y, z) = (y' + z') (x' + z)$ using truth table.

Solution:

x	y	z	$f(x, y, z) = (y' + z')(x' + z)$
0	0	0	1
0	0	1	1
0	1	0	1
0	1	1	0
1	0	0	0
1	0	1	1
1	1	0	0
1	1	1	0

Scan $f(x, y, z)$ in the table from top to bottom, if zero is encountered write the term corresponding to it in the sum of product.

$$f(x, y, z) = (x + y' + z')(x' + y + z)(x' + y' + z)(x' + y' + z)$$

6.4 SIMPLIFICATION BY KARNAUGH-MAP (K-MAP)

Karnaugh-map (K-map) is an easy method to simplify the complex Boolean function; this method will be helpful when the numbers of variables are three and more than three. The K-map is an alternative method to represent truth table. The truth table for two variables is given in Table 6.1.

122 Digital Logic Design

Let x and y are two variables which are represented in K-map as

Table 6.1 Two variable K-map

x\y	0	1
0	0	1
1	2	3

Three variable K-map: Let x, y and z are three variables which can be represented in K-map is given in Table 6.2.

Table 6.2 Three variable K-map

x\yz	00	01	11	10
0	0	1	3	2
1	4	5	7	6

OR

Table 6.3 Other three variable K-map

xy\z	0	1
00	0	1
01	2	3
11	6	7
10	4	5

Four variable K-map: Let x, y, z and w are four variables which can be represented in K-map is given in Table 6.4.

Table 6.4 Four variable K-map

xy\zw	00	01	11	10
00	0	1	3	2
01	4	5	7	6
11	12	13	15	14
10	8	9	11	10

Five variable K-map: Let x, y, z, w and v are five variables which can be represented in K-map is given in Table 6.5.

Table 6.5 Five variable K-map

xyz\vw	00	01	11	10
000	0	1	3	2
001	4	5	7	6
011	12	13	15	14
010	8	9	11	10
110	24	25	27	26
111	28	29	31	30
101	20	21	23	22
100	16	17	19	18

Six variable K-map: Let x, y, z, w, v and u are six variables which can be represented in K-map is given in Table 6.6.

Table 6.6 Six variable K-map

xyz\uvw	000	001	011	010	110	111	101	100
000	0	1	3	2	6	7	5	4
001	8	9	11	10	14	15	13	12
011	24	25	27	26	30	31	29	28
010	16	17	19	18	22	23	21	20
110	48	49	51	50	54	55	54	53
111	56	57	59	58	62	63	61	60
101	40	41	43	42	46	47	45	44
100	32	33	35	34	38	39	37	36

Algorithm for simplifying Boolean function using K-map

The following are the various steps to simplify the Boolean function using K-map.

Step 1: Place the variables in row and column as per your combination of variables.
Step 2: Put 1 in front of row and column for the value of variables in the function.
Step 3: Make the group of 2, 4, 8, 16 and so on, row-wise or column-wise, do not make group diagonal and try to accommodate maximum elements in one group.
Step 4: Write the term according to the group in the K-map, eliminate the variable from the term which value is going to change from origin to end in the group.
Step 5: Find the sum of all terms obtained in step 4 to get the final simplified Boolean function.

EXAMPLE 6.7 Simplify the following Boolean function using K-map

$$f(x, y, z) = \sum (1, 2, 3, 4, 6, 7)$$

Solution:

xy\z	0	1
00		1
01	1	1
11	1	1
10	1	

The group of the above table is given below:

xy\z	0	1
00		1
01	1	1
11	1	1
10	1	

In the above table there are three groups, so three terms will be present in the simplified function. Simplified Boolean function is $f = x'z + y + xz'$.

EXAMPLE 6.8 Simplify the following Boolean function using K-map

$$f(x, y, z, w) = \sum(1, 2, 3, 4, 6, 7, 10, 11)$$

Solution:

xy\zw	00	01	11	10
00		1	1	1
01	1		1	1
11				
10			1	1

The group of the above table is given below:

xy\zw	00	01	11	10
00		1	1	1
01	1		1	1
11				
10			1	1

In the above table there are four groups so four terms will be present in the simplified function. Simplified Boolean function is $f = x'y'w + x'z + x'yw' + y'z$.

EXAMPLE 6.9 Simplify the following Boolean function using K-map

$$f(x, y, z, w, v) = \sum(1, 2, 3, 4, 6, 7, 10, 11, 17, 18)$$

Solution:

xy\zwv	000	001	011	010	110	111	101	100
00		1	1	1	1	1		1
01			1	1				
11								
10			1	1				

The group of the above table is given below:

xy\zwv	000	001	011	010	110	111	101	100
00		1	1	1	1	1		1
01			1	1				
11								
10			1	1				

In the above table there are four groups, so four terms will be present in the simplified function.

Simplified Boolean function is $f = x'y'w + x'z'w + y'z'w + x'y'w'$.

EXAMPLE 6.10 Simplify the following Boolean function using K-map

$$f(x, y, z, w, v) = \sum(10, 11, 17, 18, 32, 33)$$

Solution:

xyu\zwv	000	001	011	010	110	111	101	100
000								
001			1	1				
011								
010		1		1				
110								
111								
101								
100	1	1						

The group of the above table is given below:

xyu\zwv	000	001	011	010	110	111	101	100
000								
001			1	1				
011								
010		1		1				
110								
111								
101								
100	1	1						

In the above table there are three groups so three terms will be present in the simplified function. The simplified function is $f(x, y, z, w, u, v) = x'y'uz'w + x'yu'z' + xy'u'z'w'$.

EXAMPLE 6.11 Simplify the following Boolean function using K-map

$$f(x, y, z, w) = \sum(1, 2, 3, 4, 6, 7, 10, 11) + \sum^{d}(5, 8, 9)$$

Solution:

xy\zw	00	01	11	10
00		1	1	1
01	1	d	1	1
11				
10	d	d	1	1

In the above K-map d represents don't care condition. The group of the above table is given below:

xy\zw	00	01	11	10
00		1	1	1
01	1	d	1	1
11				
10	d	d	1	1

In the above table there are five groups, so five terms will be present in the simplified function. The simplified function is $f(x, y, z, w) = x'w + x'z' + y'w + y'z$.

EXAMPLE 6.12 Simplify the following Boolean function using K-map

$$f(x, y, z, w) = \prod(1, 2, 3, 5, 6, 7)$$

Solution:

xy\zw	00	01	11	10
00		0	0	0
01		0	0	0
11				
10				

The group of the above table is given below:

xy\zw	00	01	11	10
00		0	0	0
01		0	0	0
11				
10				

In the above table there are two groups, so two terms in product of sum will be present in the simplified function. The simplified function is $f(x, y, z, w) = (x + w')(x + z')$.

EXAMPLE 6.13 Simplify the following Boolean function using K-map

$$f(x, y, z, w) = \prod(1, 2, 3, 5, 6, 7) + \prod^{d}(8, 9)$$

Solution:

xy\zw	00	01	11	10
00		0	0	0
01		0	0	0
11				
10	d	d		

In the above K-map d represents don't care condition. The group of the above table is given below:

xy\zw	00	01	11	10
00		0	0	0
01		0	0	0
11				
10	d	d		

In the above table there are two groups, so two terms in product of sum will be present in the simplified function. The simplified function is $f(x, y, z, w) = (x + w') (x + z')(x' + z)$.

6.5 TABULAR METHOD FOR SIMPLIFICATION

The tabulation method is also used to simplify the Boolean function. Let us consider a support set of $f: X = \{x_1, x_2, x_3, x_4, \ldots, x_n\}$ and $(x_i)_i^y$ where i is represented as:

x_i if y_i = '0'
$\quad\quad\quad x_i$ if y_i = '1'
$\quad\quad\quad 1$ if y_i = '–'

If there is no y_i = –, then you have a minterm and it can be represented by decimal equivalent of y_i. Let us consider an example, suppose a set $X = \{x_1, x_2, x_3\}$

$$(x_1)^1(x_2)^0(x_3)^1 = m_5, \text{ a minterm} \rightarrow y_1y_2y_3 = 101 = 5,$$
$$(x_1)^0(x_2)^-(x_3)^0 = a \text{ 3 cube} \rightarrow 0 - 0$$

Basic operation in tabulation method is that two (2) cubes are different at a single y_i is merged into a single cube. Let us consider an example, let $a = 1 - 1$ and $b = 0 - 1$, then merge a and b into c as $c = -1$. Star operator is also called merging and it is a special case of consensus. In tabulation method consider input and output.

Input: Let us consider a set of minterms is F.
Output: All essential prime implicants and as few prime implicants as possible are set of F.

Finding few prime implicants as possible

The Boolean function is reduced to the set covering; there is a problem for unate functions. The unate function is defined as a constant or it is represented by a sum of product function using either complemented literals for each variable. The function should be reduced to the minimum cost assignment.

Algorithm

Step 1: You are required to convert minterm list to prime implicants list.
Step 2: Firstly choose all essential prime implicants
If all minterms are covered in it then stop.
Else
Go to Step 3
Step 3: Make the reduced cover table by omitting the rows or columns.
If cover table is reduced using dominance properties,
Go to STEP 2
Else
Cyclic cover problem must solve as follows:
- Use exact method; it is also known as exponentially complex.
- Use heuristic method; it is also known as possibly non-optimal result.

Quine–McCluskey means using a "Branch and Bound" heuristic and Petrick's method is exact technique to generate all solutions allowing the best to be used.

Explanation of STEP 1

(1) You have to partition prime implicants or minterms according to the number of 1s.
(2) In this step you have to check adjacent classes for cube merging while building a new list.
(3) If you are making new entry in the list, means you are also covering entry in current list.
(4) Consider current list is CL and new list is NL, if CL = NL, stop
Else
CL ← NL
NL ← EMPTY
Go to STEP (1)

Let us consider an example. A Boolean function $f(x, y, z) = x'y'z' + xyz + x'yz + xyz'$. The minterm is on $f = \{m_0, m_3, m_6, m_7\} = \sum(0, 3, 6, 7)$ is given in Figure 6.1.

Minterm	Cube			
0	0	0	0	√
3	0	1	1	√
6	1	1	0	
7	1	1	1	√

Figure 6.1 Minterm on $f = \{m_0, m_3, m_6, m_7\}$.

The reduced minterm is given in Figure 6.2.

Minterm	Cube				
3, 7	–	1	1		√
6, 7	1	1	–		√

Figure 6.2 Reduced minterm.

$$f(x, y, z) = (-11 + 11-) = x'y'z' + yz + xy$$

EXERCISES

Short Answer Questions

1. What is canonical form?
2. What is p prime implicant?
3. What is K-map?
4. What are various methods to simplify the Boolean function?
5. Differentiate between SOP and POS.
6. Why is exchange the place of 10 and 11 in K-map?
7. Why is group of 2, 4, 8,… in K-map?
8. Why is group not for nm diagonally?

Long Answer Questions

1. Simplify the following Boolean function using Boolean algebra.
   (i) $f(x, y, z) = x'yz + xyz + xy'z' + x'$
   (ii) $f(x, y, z) = xy'z + xyz + xy' + xy$
   (iii) $f(x, y, z) = xz' + xyz + xz + xy'$
   (iv) $f(x, y, z) = xyz + xy + x'y + x'yz$
   (v) $f(x, y, z) = yz + xy + yz'$
2. Simplify the following product of sum Boolean function using Boolean algebra
   (i) $f(x, y, z) = (x' + y + z)(x + y + z)(x' + y)$
   (ii) $f(x, y, z) = (x + y' + z)(x + y + z)(x + y')$
   (iii) $f(x, y, z) = (x + y + z')(x' + y + z)(x + z')$
   (iv) $f(x, y, z) = (x + y + z')(x' + y + z)(x + y' + z')$
   (v) $f(x, y, z) = (x' + y + z')(x' + y + z')(y' + z')$
3. Simplify the following Boolean function using Boolean algebra
   (i) $f(x, y, z) = \sum(0, 2, 3, 5, 6, 7)$

(ii) $f(x, y, z, w) = \sum(0, 1, 4, 5, 7, 8, 9)$

(iii) $f(x, y, z) = \sum(2, 3, 6, 7)$

(iv) $f(x, y, z, w) = \sum(1, 2, 3, 5, 11, 12, 13)$

(v) $f(x, y) = \sum(0, 2, 3)$

4. Simplify the following Boolean function using Boolean algebra

   (i) $f(x, y, z) = \prod(1, 2, 5, 6, 7)$

   (ii) $f(x, y, z) = \prod(0, 2, 4, 5, 6, 7)$

   (iii) $f(x, y, z, w) = \prod(0, 2, 3, 5, 6, 7, 8)$

   (iv) $f(x, y, z) = \prod(1, 2, 4, 5, 6)$

   (v) $f(x, y, z) = \prod(1, 3, 6, 7)$

5. Simplify the following Boolean function using K-map

   (i) $f(x, y, z) = (x' + y + z)(x + y + z)(x' + y)$

   (ii) $f(x, y, z) = (x + y' + z)(x + y + z)(x + y')$

   (iii) $f(x, y, z) = (x + y + z')(x' + y + z)(x + z')$

   (iv) $f(x, y, z) = (x + y + z')(x' + y + z)(x + y' + z')$

   (v) $f(x, y, z) = (x' + y + z')(x' + y + z')(y' + z')$

6. Simplify the following Boolean function using K-map

   (i) $f(x, y, z) = \sum(0, 2, 3, 5, 6, 7)$

   (ii) $f(x, y, z, w) = \sum(0, 1, 4, 5, 7, 8, 9) + \sum^{d}(6, 10, 11)$

   (iii) $f(x, y, z) = \sum(2, 3, 6, 7) + \sum(1, 5)$

   (iv) $f(x, y, z, w) = \sum(1, 2, 3, 5, 11, 12, 13)$

   (v) $f(x, y) = \sum(0, 2, 3)$

   (vi) $f(x, y, z, w, u) = \sum(1, 2, 3, 5, 11, 12, 13, 16, 18)$

   (vii) $f(x, y, z, w, u, v) = \sum(1, 2, 11, 12, 13, 31, 32)$

7. Simplify the following Boolean function using K-map

   (i) $f(x, y, z) = \prod(1, 2, 5, 6, 7)$

   (ii) $f(x, y, z) = \prod(0, 2, 4, 5, 6, 7) + \prod^{d}(1, 3)$

   (iii) $f(x, y, z, w) = \prod(0, 2, 3, 5, 6, 7, 8) + \prod^{d}(1, 4, 9)$

(iv) $f(x, y, z) = \prod(1, 2, 4, 5, 6)$

(v) $f(x, y, z) = \prod(1, 3, 6, 7)$

(vi) $f(x, y, z, w, u) = \prod(0, 2, 3, 5, 6, 7, 8, 16, 17) + \prod^{d}(1, 4, 9)$

(vii) $f(x, y, z, w, u, v) = \prod(0, 2, 3, 5, 6, 7, 8, 32, 33) + \prod^{d}(1, 4, 9)$

(viii) $f(x, y, z, w, u) = \prod(0, 2, 3, 5, 6, 7, 8, 19, 20)$

(ix) $f(x, y, z, w, u, v) = \prod(0, 2, 3, 5, 6, 7, 8, 34, 35)$

8. Convert the following in canonical form
   (i) $f(x, y, z) = x'y + yz + xy' + x'$
   (ii) $f(x, y, z) = y'z + xz + xy' + xy$
   (iii) $f(x, y, z) = xz' + xy + xz + xy'$
   (iv) $f(x, y, z) = yz + xy + x'y + x'y$
   (v) $f(x, y, z) = xz + xy + yz'$

9. Convert the following in canonical form
   (i) $f(x, y, z) = (y + z)(x + y)(x' + y)$
   (ii) $f(x, y, z) = (y' + z)(x + y)(x + y')$
   (iii) $f(x, y, z) = (y + z')(x' + y + z)(x + z')$
   (iv) $f(x, y, z) = (x + z')(y + z)(x + y' + z')$
   (v) $f(x, y, z) = (x' + y)(y + z')(y' + z')$

10. Convert the following in canonical form
    (i) $f(x, y, z) = \sum(0, 2, 3, 5, 6, 7)$
    (ii) $f(x, y, z, w) = \sum(0, 1, 4, 5, 7, 8, 9)$
    (iii) $f(x, y, z) = \sum(2, 3, 6, 7)$
    (iv) $f(x, y, z, w) = \sum(1, 2, 3, 5, 11, 12, 13)$
    (v) $f(x, y) = \sum(0, 2, 3)$

Multiple Choice Questions

1. The canonical form of Boolean function $f(x, y, z) = xy + yz'x$ is
   (a) $xyz + yz'x'$
   (b) $xyz + yz'$
   (c) $xyz + xyz'$
   (d) $xyz + 2xyz'$

2. The canonical form of Boolean function $f(x, y, z) = (x + y)(y + z' + x)$ is
   (a) $(x + y + z)(y + z' + x)$
   (b) $(x + y + z)(y + z')$
   (c) $(x + y + z)(x' + y + z')$
   (d) $(x + y + z)(x + y + z')$

3. How many groups are possible in the following K-map?

xy\z	0	1
00		1
01	1	1
11	1	1
10	1	

(a) 2 (b) 3
(c) 4 (d) 5

4. The simplified function derived from the truth table mentioned in Question no. 3 is
   (a) $f = x'z + y + xz'$
   (b) $f = x'z + xy + xz'$
   (c) $f = x'z + yz + xz'$
   (d) $f = x'z + y + z'$

5. The truth table is given below, determine the sum of product.

Input			Output
x	y	z	O
0	0	0	0
0	0	1	1
0	1	0	0
0	1	1	1
1	0	0	0
1	0	1	1
1	1	0	0
1	1	1	1

(a) $O = x'y'z + x'yz + xy'z + xyz$
(b) $O = x'yz + x'yz + xy'z + xyz$
(c) $O = xy'z + x'yz + xyz$
(d) $O = x'y + x'yz + xy'z + xyz$

6. The method for reduction of digital logic circuits is:
   (a) Boolean Algebra
   (b) K-map
   (c) Symbol reduction
   (d) (a) and (b)

7. How many gates would be required to implement the following given Boolean function before simplification? $abc + b(c + d) + a(d + c) + c(c + d)$
   (a) 5 (b) 6
   (c) 7 (d) 8

8. Using K-map to simplify the Boolean expression, grouping cells within a K-map, the group of _____ .
   (a) 1, 2, 3, 4
   (b) 2, 4, 6, 8
   (c) 2, 4, 8, 16
   (d) 3, 6, 9, 12

9. The sum of product from the given K-map is

xy\zw	00	01	11	10
00		1	1	1
01	1		1	1
11				
10			1	1

(a) $xz + yz + x\,yw + x\,y\,w$
(b) $xz + yz + x\,y\,w$
(c) $xz + yz + x\,yw$
(d) $xz + yz + x\,y + x\,y\,w$

10. Mapping the sum of product function $x\,y\,z + xy\,z + xyz$ is

(a)
xy\z	0	1
00	1	
01		
11		1
10		1

(b)
xy\z	0	1
00		1
01		1
11		1
10		1

(c)
xy\z	0	1
00	1	1
01	1	
11		1
10		1

(d)
xy\z	0	1
00	1	
01		
11		1
10	1	

Answers

1. (c)   2. (d)   3. (b)   4. (a)   5. (a)   6. (d)
7. (b)   8. (c)   9. (a)   10. (a)

BIBLIOGRAPHY

http://dictionary.reference.com/browse/postulate.

http://www.proofwiki.org/wiki/Definition:Boolean_Function.

http://www.webopedia.com/TERM/L/logic_gate.html.

Whitesitt, J.E., *Boolean Algebra and Its Applications*. Dover Books on Mathematics, ISBN-13: 978-0486684833.

CHAPTER 7

Combinational Circuit Design

7.1 INTRODUCTION

The combinational circuit is a combination of various logic gates for a particular purpose. The feedback is not required in the combinational circuit. In the combinational circuit we give the input and get the output. The examples of combinational circuit are encoder, decoder, multiplexer, demultiplexer, half adder, full adder, half subtractor, full subtractor, etc. Then combinational circuit is not used for the purpose of memory. The combinational circuits have no capability to store the data. The combinational circuit uses many inputs and many outputs means it accepts n number of inputs and produce m number of outputs. The block diagram of combinational circuit is given in Figure 7.1.

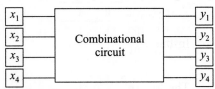

Figure 7.1 Combinational circuit design procedure.

There are several steps to design a combinational circuit that is known as the combinational circuit design procedure. You can follow steps given here to design the combinational circuit in easy ways. The following are the steps to design the combinational circuit.

Step 1: Define the problem clearly as well as understand the problem.
Step 2: Find the number of input and output required to design a circuit as per problem defined in Step 1.
Step 3: Assign the variable to all inputs and outputs, for example, let three inputs and two outputs, assign x, y and z are three input variables and a and b are two output variables.
Step 4: Find the relationship between input and output.
Step 5: Draw the truth table according to the relationship in Step 4.
Step 6: Find minimized Boolean function using K-map.
Step 7: Draw the logic diagram for the Boolean function obtained in Step 6.

EXAMPLE 7.1 Design a combinational circuit for 3-bit even parity checker.

Solution: Let x, y and z are three input variables and p is the output for parity bit.
The truth table is as follows:

Input			Output
x	y	z	p
0	0	0	0
0	0	1	1
0	1	0	1
0	1	1	0
1	0	0	1
1	0	1	0
1	1	0	0
1	1	1	1

K-map for p

xy/z	0	1
00		1
01	1	
11		1
10	1	

In this table there is no group, so the simplified Boolean function for p is

$$p = f(x, y, z) = x'y'z + x'yz' + xyz + xy'z'$$

The logic diagram is given below:

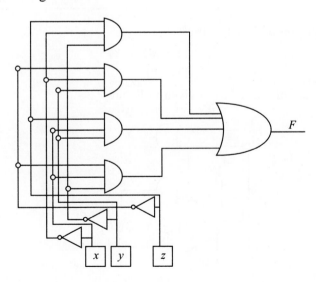

7.2 HALF ADDER

Half adder is a combinational circuit with two inputs and two outputs as sum and carry. This circuit is meant to feature 2 single bit binary number as X and Y. The block diagram of half adder is given in Figure 7.2.

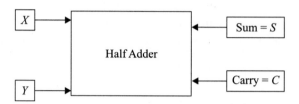

Figure 7.2 Half adder block diagram.

The truth table of half adder is given in Table 7.1

Table 7.1 Half adder truth table

Input		Output	
X	Y	S	C
0	0	0	0
0	1	1	0
1	0	1	0
1	1	0	1

The Boolean function of S and C are

$$S = f(X, Y) = X'Y + XY'$$
$$C = f(X, Y) = XY$$

The circuit diagram of half adder is given in Figure 7.3

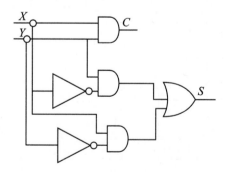

Figure 7.3 Half adder circuit diagram.

7.3 FULL ADDER

Full adder is combination circuits that overcome the drawback of half adder combinational circuit. It can accept three bit as input and produce two bit output as sum and carry. It can add three bit numbers X, Y, and Z, produce S and C as sum and carry. The full adder is a three input and two output combinational circuit.

Figure 7.4 Full adder block diagram.

The block diagram of full adder is given in Figure 7.4

The truth table of full adder is given in Table 7.2.

Table 7.2 Full adder truth table

Input			Output	
X	Y	Z	S	C
0	0	0	0	0
0	0	1	1	0
0	1	0	1	0
0	1	1	0	1
1	0	0	1	0
1	0	1	0	1
1	1	0	0	1
1	1	1	1	1

The Boolean function of sum and carry as S and C of full adder is given below:

$$S = f(X, Y, Z) = \sum(1, 2, 4, 7)$$

$$C = f(X, Y, Z) = \sum(3, 5, 6, 7)$$

K-map for S is given in Figure 7.5.

XY/Z	0	1
00		1
01	1	
11		1
10	1	

Figure 7.5 K-map for S.

$$S = f(x, y, z) = x'y'z + x'yz' + xy'z' + xyz$$

K-map for C is given in Figure 7.6.

xy/z	0	1
00		
01		1
11	1	1
10		1

Figure 7.6 K-map for C.

$$C = f(x, y, z) = yz + xy + xz$$

The circuit diagram of full adder is given in Figure 7.7

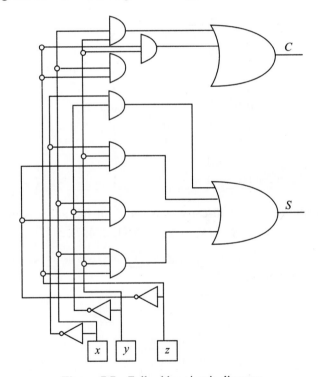

Figure 7.7 Full adder circuit diagram.

7.4 N-BIT PARALLEL ADDER

The full adder is capable of adding solely 2 single digit binary range at the side of a carry input. However in sensible we want to add binary numbers that square measure for much longer than simply one bit. To add 2 n-bit binary numbers we want to use the n-bit parallel adder. It uses many of full adders in cascade. The carry output of the previous full adder is connected to hold input of succeeding full adder.

7.4.1 Four-bit Parallel Adder

In the diagram, A_0 and B_0 represent the LSB of the four-bit words A and B. Therefore, full adder-0 is the lowest stage. Therefore, its C_{in} has been for good created zero. The remainders of the connections are unit specifically same as those of n-bit parallel adder is shown in Figure 7.8. The four-bit parallel adder may be a quite common logic circuit, which is given in the figure.

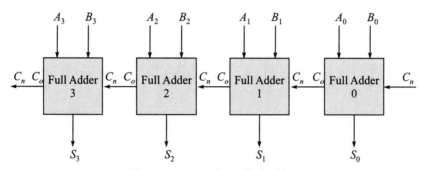

Figure 7.8 *N*-bit parallel adder.

7.5 HALF SUBTRACTOR

The half subtractor is a combinational circuit. It could be a combination circuit with two inputs and two outputs as difference and borrow. It produces the distinction between the 2 binary bits at the input and additionally produces an output (borrow) to point if a one has been borrowed. It performs subtraction $(X - Y)$, X is named as number bit and Y is named as another number bit, which is to be subtracted from X. The truth table of half subtractor is given in Table 7.3.

Table 7.3 Half subtractor truth table

Inputs		Output $(X - Y)$	
X	Y	D	B
0	0	0	0
0	1	1	1
1	0	1	0
1	1	0	1

Boolean function of half subtractor is given below. The difference is represented by D as follows.

$$D = f(X, Y) = \sum(1, 2)$$
$$D = f(X, Y) = 01 + 10$$
$$D = f(X, Y) = 01 + 10$$
$$D = f(X, Y) = X'Y + XY'$$

The borrow is represented by B as follows.

$$B = f(X, Y) = \sum(1, 3)$$
$$B = f(X, Y) = 01 + 11$$
$$B = f(X, Y) = X'Y + XY$$

The logical diagram of half subtractor is given in Figure 7.9.

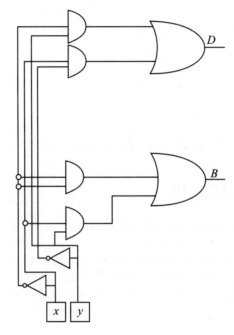

Figure 7.9 Half subtractor circuit diagram.

7.6 FULL SUBTRACTOR

Full subtractor is a combination circuit which used three inputs as X, Y, Z and produces two outputs as difference D and borrows B. The full subtractor is used to overcome the disadvantage of a half subtractor. It performed the subtraction $(X - Y - Z)$, in which X is the minuend, Y and Z are subtrahends. The truth table of full subtractor is given in Table 7.4.

Table 7.4 Full subtractor

Inputs			Output $(X - Y - Z)$	
X	Y	Z	D	B
0	0	0	0	0
0	0	1	1	1
0	1	0	1	1
0	1	1	0	1
1	0	0	1	0
1	0	1	0	0
1	1	0	0	0
1	1	1	1	1

Boolean function of half subtractor is given below.

$$D = f(X, Y, Z) = \sum(1, 2, 4, 7)$$
$$B = f(X, Y, Z) = \sum(1, 2, 3, 7)$$

K-map for D is given in Figure 7.10.

XY/Z	0	1
00		1
01	1	
11		1
10	1	

Figure 7.10 K-map for D.

$$f(x, y, z) = x'y'z + x'yz' + xy'z' + xyz$$

K-map for B is given in Figure 7.11.

XY/Z	0	1
00		1
01	1	1
11		1
10		

Figure 7.11 K-map for B.

$$B = f(x, y, z) = x'y + yz + x'z$$

The logic circuit diagram is given in Figure 7.12.

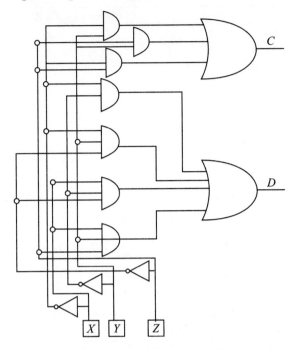

Figure 7.12 Full subtractor circuit diagram.

7.7 FOUR-BIT PARALLEL SUBTRACTOR

The number to be subtracted Y is first passed through inverters to obtain its one's complement. The 4-bit adder then adds X and two's complement of Y to produce the subtraction. $S_3\ S_2\ S_1\ S_0$ represent the result of binary subtraction $X - Y$ and carry output C represents the polarity of the result. If $X > Y$ then $C = 0$ and the result of binary form $(X - Y)$ then $C = 1$ and the result is in the two's complement form. The block diagram in Figure 7.13 shows the 4-bit parallel subtractor.

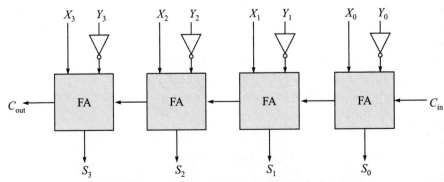

Figure 7.13 4-bit parallel subtractor block diagram.

In Figure 7.13 $X = x_3x_2x_1x_0$, $Y = y_3y_2y_1y_0$ and $-Y = y_3'y_2'y_1'y_0'$. The circuit will perform the addition as $X + (-Y)$ to give the sum as $S_3S_2S_1S_0$. This circuit will do the 4 bits subtraction.

7.7.1 N-bit Parallel Subtractor

The subtraction will be administrated by taking the one's or two's complement of the number to be deducted. For example, you are able to perform the subtraction $(X - Y)$ by adding either one's or two's complement of Y to X. This means you are able to use a binary adder to perform the binary subtraction. The N-bit subtractor is given in Figure 7.14.

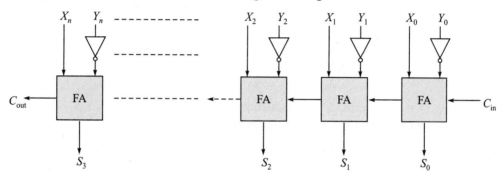

Figure 7.14 N-bit subtractor.

In Figure 7.14 $X = x_n...x_2x_1x_0$, $Y = y_n...y_2y_1y_0$ and $-Y = y_n'... y_2'y_1'y_0'$. This circuit will perform the addition $X + (-Y)$ to produce the sum as $S_n...S_2S_1S_0$. In this way the above circuit can perform the subtraction of two n-bit numbers.

7.8 MULTIPLEXERS

Multiplexer is a special form of combinational circuit. This is a digital circuit that accepts more than one input lines and only one output line. There are n input data as $I_1, I_2, ..., I_n$, out of n input data only one input data will be routed to output line. The input line is selected for output line with the help of control signal or selector line. The control signal or selector line is decided with the help of equation $2^c = n$, where c is control signal or control lines and n is input number of lines. The choice of one of the n input lines is finished by the chosen inputs. Betting on the digital code applied at the chosen inputs, one out of n information sources is chosen and transmitted to the only output Y. E is named the stroboscope or modify input that is helpful for the cascading. It is usually a vigorous low terminal, meaning it will perform the specified operation once when it is zero. The block diagram of multiplexer is given in Figure 7.15.

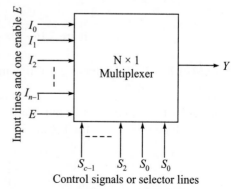

Figure 7.15 N × 1 multiplexer.

7.8.1 4 × 1 Multiplexer

The 4 × 1 multiplexer has four input lines and one output line. The control signal or selector line is decided by the equation $2^c = n$, in 4 × 1 multiplexer input is 4, so control signal or selector lines are $2^c = 4$. The value of c is determined by $2^c = 2^2$. In this equation base is same for both sides, so compare power, then $c = 2$. Now control signals or selector line is two (2) for 4 × 1 multiplexer. This two control signals or selector line will decide which input line out of four input lines will be routed to output line. Let us consider S_0 and S_1 are two control signals or selector lines, the combination of these two control signals or selector lines is the input line as follows.

$$S_0' S_1' = I_0$$
$$S_0' S_1 = I_1$$
$$S_0 S_1' = I_2$$
$$S_0 S_1 = I_3$$

The output O is represented in Boolean equation $O = S_0' S_1' I_0 + S_0' S_1 I_1 + S_0 S_1' I_2 + S_0 S_1 I_3$
The logical diagram of 4 × 1 multiplexer is shown in Figure 7.16.

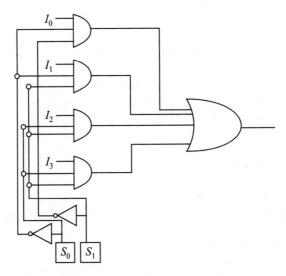

Figure 7.16 4 × 1 multiplexer logical diagram.

EXAMPLE 7.2 Implement the following Boolean function with 4 × 1 multiplexer $f(x, y, z) = \sum(0, 2, 3, 5, 6, 7)$

Solution: Let us consider x and y are control signals. Now determine the value of I_0, I_1, I_2 and I_3 as inputs of 4 × 1 multiplexer.

	I_0	I_1	I_2	I_3
z'	⓪	1	②	③
z	4	⑤	⑥	⑦

$I_0 = z'$, $I_1 = z$, $I_2 = z' + z = 1$, and $I_3 = z' + z = 1$

Now draw the block diagram (Implementation of function $f(x, y, z) = \sum(0, 2, 3, 5, 6, 7)$.

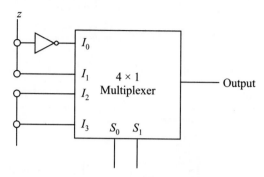

EXAMPLE 7.3 Implement the following Boolean function with 4 × 1 multiplexer $f(x, y, z) = \sum(0, 2, 3, 7)$.

Solution: Let us consider y and z are control signals. Now determine the value of I_0, I_1, I_2 and I_3 as inputs of 4 × 1 multiplexer.

	I_0	I_1	I_2	I_3
x'	(0)	1	(2)	(3)
x	4	5	6	(7)

$I_0 = x'$, $I_1 = 0$, $I_2 = x'$, and $I_3 = x' + x = 1$

Now draw the block diagram (Implementation of function $f(x, y, z) = \sum(0, 2, 3, 7)$.

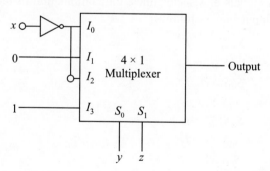

EXAMPLE 7.4 Implement the following Boolean function with 4 × 1 multiplexer $f(x, y, z) = \sum(1, 2, 3, 6)$.

Solution: Let us consider x and z are control signals. Now determine the value of I_0, I_1, I_2 and I_3, as inputs of 4 × 1 multiplexer.

	I_0	I_1	I_2	I_3
y'	0	①	②	③
y	4	5	⑥	7

$I_0 = 0$, $I_1 = y'$, $I_2 = y' + y = 1$, and $I_3 = y'$

Now draw the block diagram

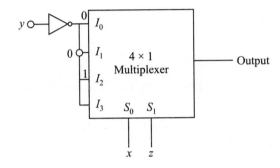

7.8.2 8 × 1 Multiplexer

The 8 × 1 multiplexer having eight input lines and one output line. The control signal or selector line is decided by the equation $2^c = n$, in 8 × 1 multiplexer input is 8, so control signal or selector lines are $2^c = 8$. The value of c is determined by $2^c = 2^3$. In this equation base is same for both sides, so compare power, then $c = 3$. Now control signals or selector line is three (3) for 8 × 1 multiplexer. This three control signals or selector line will decide which input line out of eight input lines will be routed to output line. Let us consider S_0, S_1 and S_2 are three control signals or selector lines, the combination of these three control signal or selector lines is the input line as follows:

$$S_0' S_1' S_2' = I_0$$
$$S_0' S_1' S_2 = I_1$$
$$S_0' S_1 S_2' = I_2$$
$$S_0' S_1 S_2 = I_3$$
$$S_0 S_1' S_2' = I_4$$
$$S_0 S_1' S_2 = I_5$$
$$S_0 S_1 S_2' = I_6$$
$$S_0 S_1 S_2 = I_7$$

The output O is represented in Boolean equation

$$O = S_0' S_1' S_2' I_0 + S_0' S_1' S_2 I_1 + S_0' S_1 S_2' I_2 + S_0' S_1 S_2 I_3 + S_0 S_1' S_2' I_4 + S_0 S_1' S_2 I_5 + S_0 S_1 S_2' I_6 + S_0 S_1 S_2 I_7$$

The logical diagram of 8 × 1 multiplexer is shown in Figure 7.17.

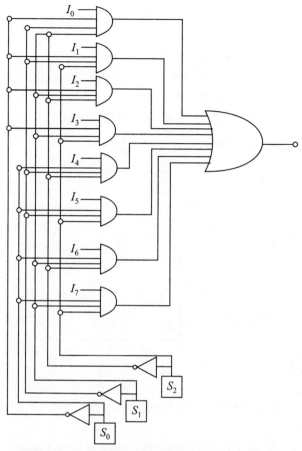

Figure 7.17 8 × 1 multiplexer logical diagram.

EXAMPLE 7.5 Implement the following Boolean function with 8 × 1 multiplexer $f(x, y, z, w) = \sum(0, 2, 3, 5, 6, 7, 8, 10, 11)$.

Solution: Let us consider x, y and z are control signals. Now determine the value of I_0, I_1, I_2, I_3, I_4, I_5, I_6, and I_7 as inputs of 8 × 1 multiplexer.

	I_0	I_1	I_2	I_3	I_4	I_5	I_6	I_7
w'	⓪	1	②	③	4	⑤	⑥	⑦
w	⑧	9	⑩	⑪	12	13	14	15

$$I_0 = w' + w = 1$$
$$I_1 = 0$$
$$I_2 = w' + w = 1$$

$I_3 = w' + w = 1$
$I_4 = 0$
$I_5 = w'$
$I_6 = w'$
$I_7 = w'$

Now draw the block diagram for implementation of function. The below figure shows the implementation of Boolean function $f(x, y, z, w) = \sum(0, 2, 3, 5, 6, 7, 8, 10, 11)$.

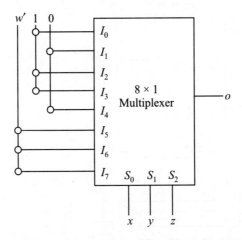

EXAMPLE 7.6 Implement the following Boolean function with 8×1 multiplexer $f(x, y, z, w) = \sum(1, 2, 3, 5, 7, 8, 11)$.

Solution: Let us consider x, y and z are control signals. Now determine the value of I_0, I_1, I_2, I_3, I_4, I_5, I_6 and I_7 as inputs of 8×1 multiplexer.

	I_0	I_1	I_2	I_3	I_4	I_5	I_6	I_7
w'	0	①	②	③	4	⑤	6	⑦
w	⑧	9	10	⑪	12	13	14	15

$I_0 = w$
$I_1 = w'$
$I_2 = w'$
$I_3 = w' + w = 1$
$I_4 = 0$
$I_5 = w'$
$I_6 = 0$
$I_7 = w'$

Now draw the block diagram for implementation of function. The below figure shows the implementation of Boolean function $f(x, y, z, w) = \sum(1, 2, 3, 5, 7, 8, 11)$.

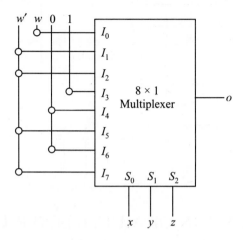

EXAMPLE 7.7 Implement the following Boolean function with 8 × 1 multiplexer $f(x, y, z, w) = \prod(1, 2, 3, 5, 7, 8, 11)$.

Solution: Let us consider x, y and z are control signals. Now determine the value of I_0, I_1, I_2, I_3, I_4, I_5, I_6 and I_7 as inputs of 8 × 1 multiplexer.

	I_0	I_1	I_2	I_3	I_4	I_5	I_6	I_7
w'	0	(1)	(2)	(3)	4	(5)	6	(7)
w	(8)	9	10	(11)	12	13	14	15

$$I_0 = w'$$
$$I_1 = w$$
$$I_2 = w$$
$$I_3 = (w) \times (w') = 0$$
$$I_4 = 1$$
$$I_5 = w$$
$$I_6 = 1$$
$$I_7 = w$$

Implementation of function $f(x, y, z, w) = \prod(1, 2, 3, 5, 7, 8, 11)$ is given in the below figure.

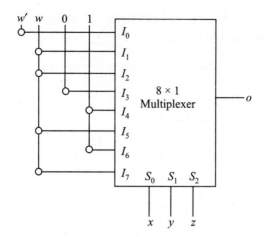

7.9 IMPLEMENTATION OF MULTIPLEXER USING K-MAP

The different types of multiplexer are also implemented using K-map. The Karnaugh map provides a simple and straightforward method for implementing the Boolean function. The input I_0, I_1, I_2, I_3, I_4, I_5, I_6, ...and I_n as inputs of $(n + 1) \times 1$ multiplexer is consider any side of K-map. Let us consider an example for implementation of 4×1 multiplexer, in this example you could have choose any variable to be the data variable and the other two as control signal variables. Assume one is to take z as the data variable and x, y are control signals. The corresponding Karnaugh map is given in Figure 7.18.

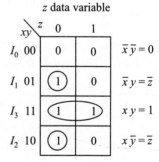

Figure 7.18 K-map with control signals xy.

The value of inputs of 4×1 multiplexer I_0, I_1, I_2, I_3 are determined from the above K-map as

$I_0 = \overline{xy} = 0$, $I_1 = \overline{x}y = \overline{z}$, $I_2 = xy = 1$, $I_3 = x\overline{y} = \overline{z}$,

The implementation of Boolean function is given in Figure 7.19.

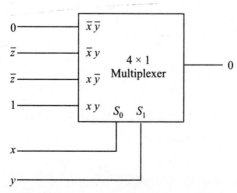

Figure 7.19 Boolean function implementation using 4×1 multiplexer.

The four variables function is implemented with 4 × 1 multiplexer with different combination of any two variables out of four variables as control signals are given in Figure 7.20 and Figure 7.21.

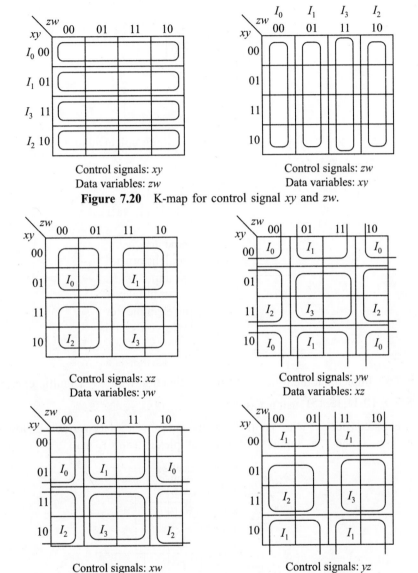

Figure 7.20 K-map for control signal *xy* and *zw*.

Figure 7.21 K-map for control signal *xz*, *yw*, *xw* and *yz*.

EXAMPLE 7.8 Implement the following Boolean function with 4 × 1 multiplexer $f(x, y, z) = \sum(1, 2, 3, 5, 7)$ using K-map.

Solution: Let us consider x and y are control signals and z is data variable. Now determine the value of I_0, I_1, I_2 and I_3 as inputs of 4 × 1 multiplexer using K-map.

152 Digital Logic Design

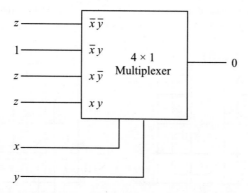

$I_0 = \overline{xy} = z$, $I_1 = \overline{x}y = 1$, $I_2 = x\overline{y} = z$, $I_3 = xy = z$, the implementation of Boolean function $f(x, y, z) = \sum(1, 2, 3, 5, 7)$ is given below:

EXAMPLE 7.9 Implement the following Boolean function with 4×1 multiplexer $f(x, y, z, w) = \sum(0, 2, 5, 7, 8, 10, 11, 13, 15)$ using K-map.

Solution: Let us consider x, y are control signals and z, w are data variables. Now determine the value of I_0, I_1, I_2 and I_3 as inputs of 4×1 multiplexer using K-map.

$I_0 = \overline{xy} = \overline{w}$, $I_1 = \overline{x}y = w$, $I_2 = x\overline{y} = w$, $I_3 = xy = \overline{w}$, the implementation of four variables using Boolean function $f(x, y, z, w) = \sum(0, 2, 5, 7, 8, 10, 11, 13, 15)$ is implemented by 4×1 multiplexer and is given as.

Combinational Circuit Design **153**

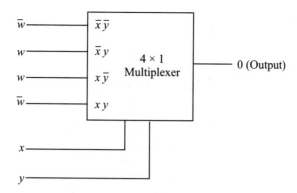

EXAMPLE 7.10 Implement the following Boolean function with 4 × 1 multiplexer $f(x, y, z, w)$ = $\sum(0, 2, 4, 5, 10, 11, 13, 15)$ using K-map.

Solution: Let us consider x, y are control signals and z, w are data variables. Now determine the value of I_0, I_1, I_2 and I_3 as inputs of 4 × 1 multiplexer using K-map.

xy\zw	00	01	11	10
00	1			1
01	1	1		
11		1	1	
10			1	1

Control signals: xy
Data variables: zw

$I_0 = \overline{xy} = \overline{w}$, $I_1 = \overline{x}y = \overline{z}$, $I_2 = x\overline{y} = z$, $I_3 = xy = w$, the implementation of four variables using Boolean function $f(x, y, z, w) = \sum(0, 2, 4, 5, 10, 11, 13, 15)$ is implemented by 4 × 1 multiplexer and is given below.

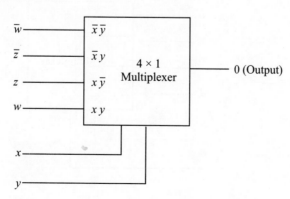

EXAMPLE 7.11 Implement the following Boolean function with 4 × 1 multiplexer $f(x, y, z, w) = xy + yz + xzw$.

Solution: The $f(x, y, z, w) = xy + yz + xzw$ is given in simplified form. Write K-map for the above simplified Boolean function as follows:

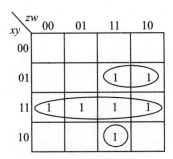

Now determine the value of I_0, I_1, I_2 and I_3 from the above K-map. Let xy are control signals and zw are data variables in the given K-map.

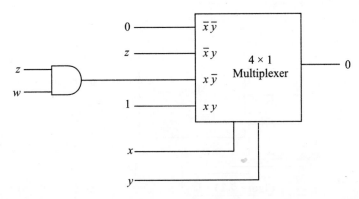

Control signals: xy
Data variables: zw

$I_0 = \overline{x}\overline{y} = 0$, $I_1 = \overline{x}y = z$, $I_2 = x\overline{y} = zw$, $I_3 = xy = 1$, the implementation of four variables using Boolean function $f(x, y, z, w) = xy + yz + xzw$ is implemented by 4 × 1 multiplexer and is given below.

7.10 DEMULTIPLEXER

Demultiplexer is a combinational circuit that performs the reverse operation of multiplexer. It accepts one input and distributes it over many output lines. Suppose n is output lines then input line is directed to one output line with the help of select lines or control signal. The control signals are determining with the help of equation $n = 2^m$, where n is the number of output line and m is the number of control signals. The block diagram of demultiplexer is given in Figure 7.22.

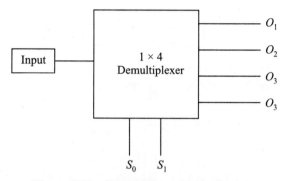

Figure 7.22 Demultiplexer block diagram.

There are various types of multiplexer which can be designed according to the requirement like 1×2, 1×4, 1×8, 1×8, 1×16 and 1×31 demultiplexer. In other words you can say $m \times n$ demultiplexer can be designed, where m represents input and n represents output lines.

7.11 DECODER

A decoder is a combinational circuit which has n input and maximum 2^n or less than 2^n outputs. Decoder is similar to a demultiplexer without having any control signals. A decoder can be represented as $I \times O$, where I represents input and O represents output, the relationship between I and O is $O \leq 2^I$. For example, consider $I = 2$, then $O = 2^2 = 4$, means 2×4 decoder. The block diagram of $I \times O$ is given in Figure 7.23.

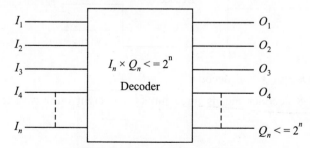

Figure 7.23 Decoder block diagram.

EXAMPLE 7.12 Design 2 × 4 decoder.

Solution: 2 × 4 decoder means two inputs and four outputs. The truth table of this decoder is given below.

Input		Outputs			
x	y	D_0	D_1	D_2	D_3
0	0	1	0	0	0
0	1	0	1	0	0
1	0	0	0	1	0
1	1	0	0	0	1

$D_0 = x'y'$, $D_1 = x'y$, $D_2 = xy'$, $D_3 = xy$
The logical diagram of 2 × 4 decoder is given below.

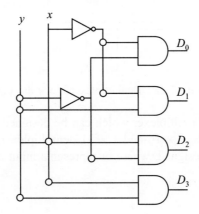

The block diagram is given below.

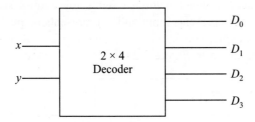

EXAMPLE 7.13 Design 3 × 8 decoder.

Solution: The 3 × 8 decoder means three inputs and eight outputs. The truth table of this decoder is given.

Combinational Circuit Design

Input			Output							
x	y	z	D_0	D_1	D_2	D_3	D_4	D_5	D_6	D_7
0	0	0	1	0	0	0	0	0	0	0
0	0	1	0	1	0	0	0	0	0	0
0	1	0	0	0	1	0	0	0	0	0
0	1	1	0	0	0	1	0	0	0	0
1	0	0	0	0	0	0	1	0	0	0
1	0	1	0	0	0	0	0	1	0	0
1	1	0	0	0	0	0	0	0	1	0
1	1	1	0	0	0	0	0	0	0	1

$D_0 = x'y'z'$, $D_1 = x'y'z$, $D_2 = x'yz'$, $D_3 = x'yz$, $D_4 = xy'z'$, $D_5 = xy'z$, $D_6 = xyz'$, $D_7 = xyz'$.

The logical diagram of 3 × 8 decoder is given below.

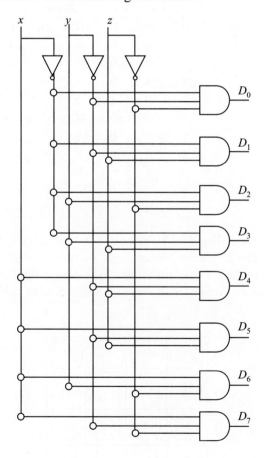

The block diagram is given below.

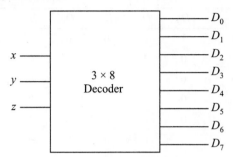

EXAMPLE 7.14 Design 4 × 10 or BCD or decimal decoder.

Solution: The 4 × 10 decoder is also known as BCD decoder or decimal decoder. This type of decoder has four inputs and ten outputs. The truth table of this decoder is given below.

Inputs				Outputs									
x	y	z	w	D_0	D_1	D_2	D_3	D_4	D_5	D_6	D_7	D_8	D_9
0	0	0	0	1	0	0	0	0	0	0	0	0	0
0	0	0	1	0	1	0	0	0	0	0	0	0	0
0	0	1	0	0	0	1	0	0	0	0	0	0	0
0	0	1	1	0	0	0	1	0	0	0	0	0	0
0	1	0	0	0	0	0	0	1	0	0	0	0	0
0	1	0	1	0	0	0	0	0	1	0	0	0	0
0	1	1	0	0	0	0	0	0	0	1	0	0	0
0	1	1	1	0	0	0	0	0	0	0	1	0	0
1	0	0	0	0	0	0	0	0	0	0	0	1	0
1	0	0	1	0	0	0	0	0	0	0	0	0	1

The input is four so, there should be 16 outputs, but outputs are ten only because the BCD goes from 0 to 9(1001). The remaining binary digits 1010, 1011, 1100, 1101 and 1111 will be used as do not care. These do not care participate in making group for finding the ten output equations. The K-map for all outputs is shown in a single K-map below.

xy \ zw	00	01	11	10
00	D_0	D_1	D_2	D_3
01	D_4	D_5	D_7	D_6
11	x	x	x	x
10	D_8	D_9	x	x

$D_0 = x'y'z'w'$, $D_1 = x'y'z'w$, $D_2 = x'y'zw'$, $D_3 = x'y'zw$, $D_4 = yz'w'$, $D_5 = yz'w$, $D_6 = yzw'$, $D_7 = yzw$, $D_8 = xw'$, $D_9 = xw$.

The logical diagram of 4 × 10 decoder is given below.

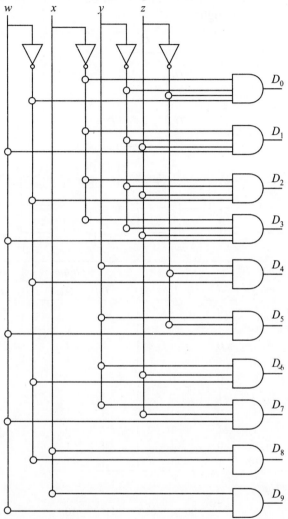

The block diagram is given below.

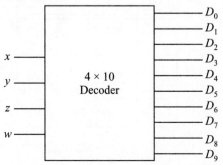

EXAMPLE 7.15 Design a seven segment decoder.

Solution: The seven segment display count number is from 0 to 9. The minimum bit required to represent 9 is four, so four bits will be inputs and outputs will be seven (7). The outputs will be determined by the given figure.

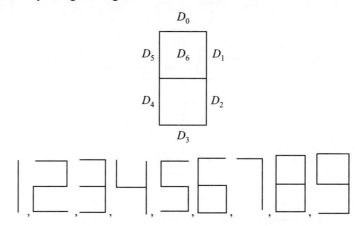

Input				Output						
x	y	z	w	D_0	D_1	D_2	D_3	D_4	D_5	D_6
0	0	0	0	1	1	1	1	1	1	0
0	0	0	1	1	1	0	0	0	0	0
0	0	1	0	1	1	0	1	1	0	1
0	0	1	1	1	1	1	1	0	0	1
0	1	0	0	0	1	1	0	0	1	1
0	1	0	1	0	1	1	1	0	1	1
0	1	1	0	1	0	1	1	1	1	1
0	1	1	1	1	1	1	0	0	0	0
1	0	0	0	1	1	1	1	1	1	1
1	0	0	1	1	1	1	1	0	1	1

K-map for D_0

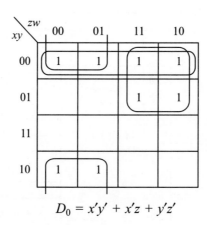

$$D_0 = x'y' + x'z + y'z'$$

Combinational Circuit Design **161**

K-map for D_1

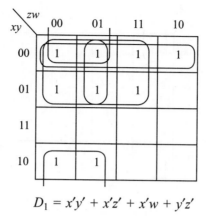

$D_1 = x'y' + x'z' + x'w + y'z'$

K-map for D_2

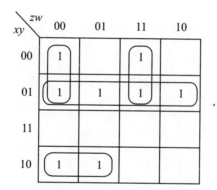

$D_2 = x'y + x'z' + x'z'w' + x'zw + xy'z'$

K-map for D_3

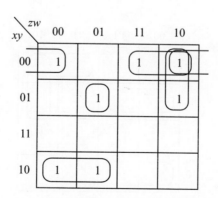

$D_3 = x'y'w' + x'y'z + x'zw' + x'yz'w + xy'z'$

K-map for D_4

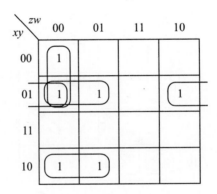

$$D_4 = y'z'w' + x'zw'$$

K-map for D_5

$$D_5 = x'z'w' + x'yz' + xy'z' + x'yw'$$

K-map for D_6

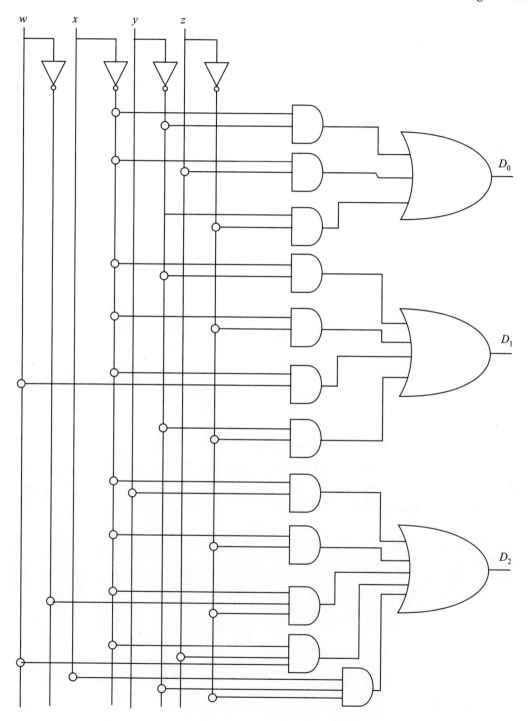

$$D_6 = x'y'z' + xy'z' + x'y'z + x'yw'$$

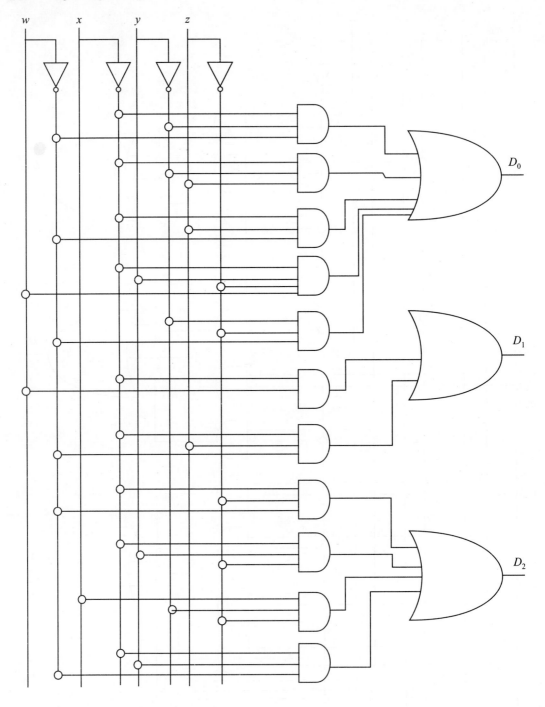

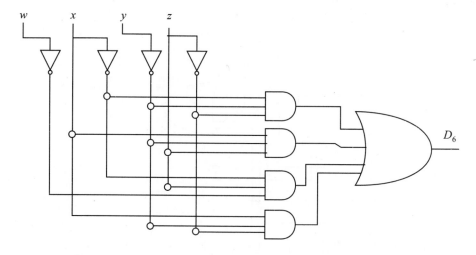

7.12 ENCODER

Encoder is a combinational circuit which reverses of decoder circuit. An encoder accepts x^n and less than its number of input lines and produces n number of output lines. The general diagram of encoder is given in Figure 7.24.

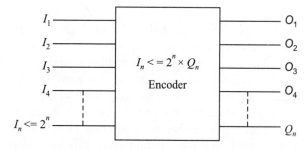

Figure 7.24 Encoder general diagram.

EXAMPLE 7.16 Design 8×3 encoder.

Solution: In this encoder there are eight input lines and three output lines. The truth table is given below.

Input								Output		
I_0	I_1	I_2	I_3	I_4	I_5	I_6	I_7	x	y	z
1	0	0	0	0	0	0	0	0	0	0
0	1	0	0	0	0	0	0	0	0	1
0	0	1	0	0	0	0	0	0	1	0
0	0	0	1	0	0	0	0	0	1	1
0	0	0	0	1	0	0	0	1	0	0
0	0	0	0	0	1	0	0	1	0	1
0	0	0	0	0	0	1	0	1	1	0
0	0	0	0	0	0	0	1	1	1	1

$$x = I_4 + I_5 + I_6 + I_7$$
$$y = I_2 + I_3 + I_6 + I_7$$
$$z = I_1 + I_3 + I_5 + I_7$$

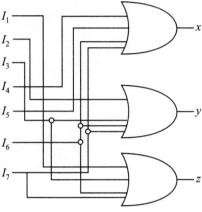

7.13 MAGNITUDE COMPARATOR

The magnitude comparator is a combinational circuit, which compares individual bit of two numbers. Let x and y are two numbers, the magnitude comparator determines whether $x > y$ or $x < y$ or $x = y$. Consider x and y are n bits number, therefore two n bits number will be input and three output lines will be greater than (GT), less than (LT) and equal to (EQ). Only one output line will be active among three output lines. If GT = 1 the LT = EQ = 0 or LT = 1 then GT = EQ = 0 or EQ = 1 then GT = LT = 0. The n bit magnitude comparator is given in Figure 7.25

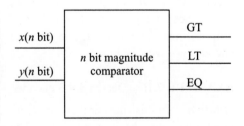

Figure 7.25 N-bit magnitude comparator.

EXAMPLE 7.17 Design a magnitude comparator for comparing two numbers of two bits.

Solution: Let X and Y are two numbers of two bits as $X = x_1 x_0$ and $Y = y_1 y_0$

Total input = 4(X = 2, Y = 2), so total possible combination is $2^4 = 16$. Making K-map is complex for 16 variables. This can be solved with conventional method.

EQ designing

Let x_i and y_i are two numbers of two bits where $i = 0, 1$.
Now define $C_i = X_i$ **XNOR** $Y_i = X_i Y_i + X_i' Y_i'$
If $C_i = 1$ then $X_i = Y_i$ for all $i = 0, 1$
If $C_i = 0$ then $X_i \neq Y_i$ for all $i = 0, 1$. So the condition for $X = Y$ or EQ = 1.
If $x_1 = y_1 \rightarrow (C_1 = 1)$ and $x_0 = y_0 \rightarrow (C_0 = 1)$, therefore EQ = $C_1 C_0$

GT designing

The output of magnitude comparator GT = 1 when $x > y$, the value of $x_1 = 1$ and $y_1 = 0$ then $x_1 > y_1$.
If $x_1 = y_1$ and $x_0 > y_0$, therefore GT = $x_1 y_1' + C_1 x_0 y_0'$.

LT designing

The output of magnitude comparator LT = 1 when $x < y$, the value of $x_1 = 0$ and $y_1 = 1$ then $x_1 > y_1$. If $x_1 = y_1$ and $x_0 < y_0$, then LT = $x_1' y_1 + C_1 x_0' y_0$.

The Boolean equation of GT, LT and EQ are
$$EQ = C_1 C_0$$
$$GT = x_1 y_1' + C_1 x_0 y_0'$$
$$LT = x_1' y_1 + C_1 x_0' y_0$$

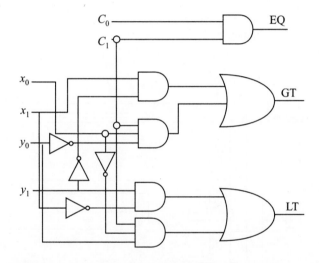

EXAMPLE 7.18 Design a magnitude comparator for comparing two numbers of four bits.

Solution: Let X and Y are two numbers of four bits as $X = x_3 x_2 x_1 x_0$ and $Y = y_3 y_2 y_1 y_0$

Total input = 8($X = 4$, $Y = 4$), so total possible combination is $2^6 = 256$. Making K-map is complex for 256 variables. This can be solved with conventional method.

EQ designing

Let x_i and y_i are two numbers of four bits where $i = 0, 1, 2, 3$.
Now define $C_i = X_i$ **XNOR** $Y_i = X_i Y_i + X_i' Y_i'$.
If $C_i = 1$ then $X_i = Y_i$ for all $i = 0, 1, 2, 3$.
If $C_i = 0$ then $X_i \neq Y_i$ for all $i = 0, 1$. So the condition for $X = Y$ or EQ = 1.
If
$$x_3 = y_3 \rightarrow (C_3 = 1) \text{ and}$$
$$x_2 = y_2 \rightarrow (C_2 = 1) \text{ and}$$
$$x_1 = y_1 \rightarrow (C_1 = 1) \text{ and}$$
$$x_0 = y_0 \rightarrow (C_0 = 1), \text{ therefore EQ} = C_3 C_2 C_1 C_0$$

GT designing

The output of magnitude comparator GT = 1 when $X > Y$, the value of $x_3 = 1$ and $y_3 = 0$ then $x_3 > y_3$.

If $x_3 = y_3$ and $x_2 > y_2$ when $x_2 = 1$, $y_2 = 0$
If $x_2 = y_2$ and $x_1 > y_1$ when $x_1 = 1$, $y_1 = 0$
If $x_1 = y_1$ and $x_0 > y_0$ when $x_0 = 1$, $y_0 = 0$
Therefore GT = $x_3 y_3' + C_3 x_2 y_2' + C_3 C_2 x_1 y_1' + C_3 C_2 C_1 x_0 y_0'$

LT designing

The output of magnitude comparator LT = 1 when $X < Y$, the value of $x_3 = 0$ and $y_3 = 1$ then $x_3 < y_3$.

If $x_3 = y_3$ and $x_2 < y_2$, when $x_2 = 0$, $y_2 = 1$
If $x_2 = y_2$ and $x_1 < y_1$, when $x_1 = 0$, $y_1 = 1$
If $x_1 = y_1$ and $x_0 < y_0$, when $x_0 = 0$, $y_0 = 1$
Therefore, GT = $x_3' y_3 + C_3 x_2' y_2 + C_3 C_2 x_1' y_1 + C_3 C_2 C_1 x_0' y_0$

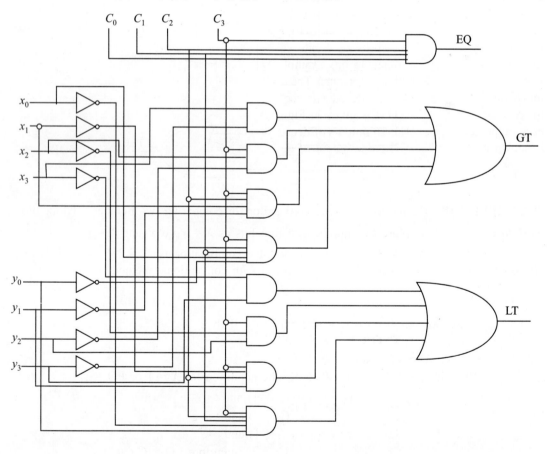

The Boolean equation of GT, LT and EQ are
  EQ = $C_3 C_2 C_1 C_0$
  GT = $x_3 y_3' + C_3 x_2 y_2' + C_3 C_2 x_1 y_1' + C_3 C_2 C_1 x_0 y_0'$
  LT = $x_3' y_3 + C_3 x_2' y_2 + C_3 C_2 x_1' y_1 + C_3 C_2 C_1 x_0' y_0$

7.14 READ ONLY MEMORY (ROM)

ROM is an important memory device which stored the information permanently. The block diagram of ROM is given in Figure 7.26. This block diagram of ROM is consisted with m inputs and n outputs. The memory address is provided by inputs. The outputs give the stored information at the location given by

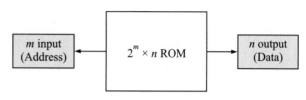

Figure 7.26 Block diagram of ROM.

address line. The information stored in the form of word. The number of word is determined with the equation "word = 2^m", where m is input. The ROM device does not have data input due to unavailability of write operation. Let us consider an example, 16 × 8 ROM. This ROM memory consists with 16 words of 4 bits each. The number of input is determining with the formula 2^m, m is the input. In this example the words can be 16 and is represented as 16 = 2^4. Therefore, the inputs are 4 (four). The four inputs form the binary number are from 0 to 16 for address. The 4 × 16 decoder is required to design this ROM. The construction of internal logic of 16 × 4 ROM is given in Figure 7.27. The memory address is represented by each output of decoder. The 16 outputs of decoder is connected with each four OR gates, which is shown in Figure 7.27. The one output of decoder is connected one input of each OR gate.

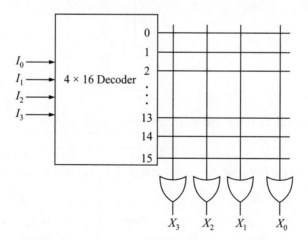

Figure 7.27 16 × 4 ROM internal logical design.

EXAMPLE 7.19 Program the ROM with the following table.

Inputs				Outputs			
I_3	I_2	I_1	I_0	X_3	X_2	X_1	X_0
0	0	0	0	1	0	1	0
0	0	0	1	1	0	1	1
0	0	1	0	0	1	1	0
0	0	1	1	0	0	1	1
0	1	0	0	1	1	1	0
⋮				⋮			
1	1	0	0	1	1	1	0
1	1	0	1	0	0	1	1
1	1	1	0	0	0	1	1
1	1	1	1	1	0	0	0

Solution:

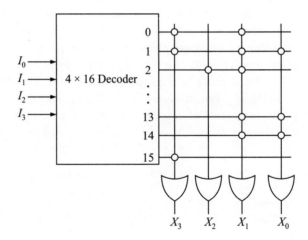

7.14.1 Combinational Circuit Implementation with ROM

The combinational circuit is implemented with ROM by choosing connection for the minterm which is also included in Boolean function. The ROM outputs are programmed to represent the Boolean functions of output variable in combinational circuit. The ROM internal operation is interpreted in two ways. The memory unit contains fixed pattern of stored word is the first interpretation. Unit implementation of the combinational circuit is second interpretation. The all output terminal is considered separately as output of Boolean function which is represented in sum of minterms as given below:

$$X_3(I_3, I_2, I_1, I_0) = \sum(2,4,6,8,...,15)$$

Combinational Circuit Design

EXAMPLE 7.20 Using ROM design a combination circuit that accepts three bits input and produces output in Excess-3 code.

Solution: The highest number in three inputs is 7, representation of 7 in Excess-3 code is 10. The 10 is represented in binary required minimum 4 bits. Therefore, the outputs will be 4 (four).

Let x, y and z are three inputs and four outputs are A, B, C and D. The truth table is given below:

Inputs			Outputs			
x	y	z	A	B	C	D
0	0	0	0	0	1	1
0	0	1	0	1	0	0
0	1	0	0	1	0	1
0	1	1	0	1	1	0
1	0	0	0	1	1	1
1	0	1	1	0	0	0
1	1	0	1	0	0	1
1	1	1	1	0	1	0

In the above truth table, any output column is not similar to input column and any output column is not zero. Therefore, the ROM truth table will be same as the above table. The same column values in input and output are that columns are removed from ROM truth table as well as the all zero value columns in outputs are also removed to get ROM truth table. The ROM truth table is given below:

Inputs			Outputs			
x	y	z	A	B	C	D
0	0	0	0	0	1	1
0	0	1	0	1	0	0
0	1	0	0	1	0	1
0	1	1	0	1	1	0
1	0	0	0	1	1	1
1	0	1	1	0	0	0
1	1	0	1	0	0	1
1	1	1	1	0	1	0

There are eight combination in inputs and four combination in outputs, therefore 8×4 decoder is used to design the circuit. The implementation circuit with ROM is given in the figure below:

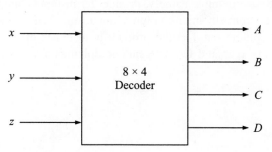

7.15 PROGRAMMABLE LOGIC ARRAY (PLA)

The concept of programmable ROM (PROM) is used in PLA (Programmable Logic Array). The concept of PROM and PLA is similar. The PLA is not full decoding of variables as well as not generating all minterms. The array of AND gates replaced decoder circuit that can be programmed to generate any product term of input variables. To provide the sum of product for the required Boolean functions the product terms are connected to OR gates. The three inputs and two outputs PLA internal logic are given in Figure 7.28 for Boolean functions $F1(x, y, z) = \sum(2, 3, 5)$ and $F2(x, y, z) = \sum(1, 4, 6)$. These two functions can be written as

$$F1 = x'yz' + x'yz + xy'z = x'y(z' + z) + xy'z = x'y + xy'z$$
$$F2 = x'y'z + xy'z' + xyz' = x'y'z + xz'(y' + y) = x'y'z + xz'$$

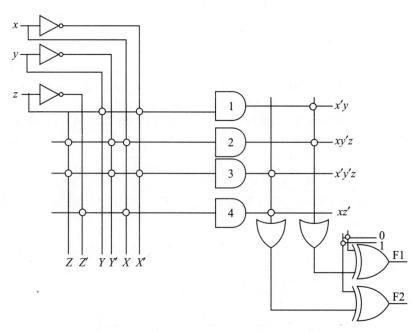

Figure 7.28 PLA internal logical diagram.

The PLA programming table is given in Table 7.3 for Figure 7.28. The PLA programming table is consisted with three columns, namely, Product term, Inputs and Output. Put all product terms in product term column. In inputs column put 0 and 1 if connection exists otherwise put dash (–). In output column put 1 if the product term exists in the function otherwise put dash (–). The T represents True and C represents complements.

Combinational Circuit Design

Table 7.3 PLA programming table

Product term	Inputs			Outputs	
				T	C
	x	y	z	F1	F2
$x'y$	0	1	–	1	–
$xy'z$	1	0	1	1	–
$x'y'z$	0	0	0	–	1
xz'	1	–	0	–	1

EXAMPLE 7.21 Implement the following Boolean function with PLA

$$F1 = x'y + xz'$$
$$F2 = x'y + xy'z$$

Solution: The given function is already simplified. The three distinct product of terms are $x'y$, xz' and $xy'z$. The PLA programming table is given below.

Product term	Inputs			Outputs	
				T	C
	x	y	z	F1	F2
$x'y$	0	1	–	1	1
xz'	1	–	0	1	–
$xy'z$	1	0	1	–	1

The PLA internal logic to implement this circuit is given below.

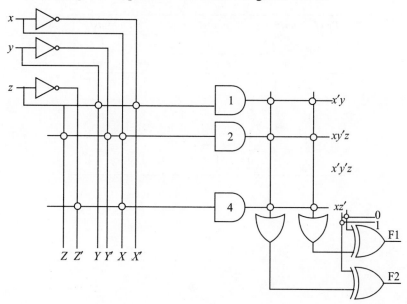

EXAMPLE 7.22 Implement the following Boolean function with PLA

$$F(x, y, z) = \sum(0, 1, 2, 4, 5, 6)$$
$$G(x, y, z) = \sum(0, 2, 3, 4, 5, 6)$$

Solution: Simplify both the functions using K-map

K-map for Boolean function $F(x, y, z) = \sum(0, 1, 2, 4, 5, 6)$

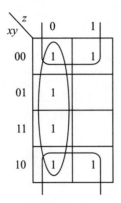 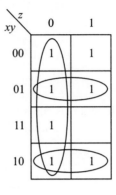

$F(x, y, z) = \sum(0, 1, 2, 4, 5, 6)\ G(x, y, z) = \sum(0, 2, 3, 4, 5, 6)$

$F(x, y, z) = z' + y'\ G(x, y, z) = z' + xy' + x'y$

There are four distinct product terms in both functions after simplification, namely, z', y', xy' and $x'y$. The PLA programming table is given below.

Product term	Inputs			Outputs	
				T	C
	x	y	z	F1	F2
z'	–	–	0	1	1
y'	–	0	–	1	–
xy'	1	0	–	–	1
$x'y$	0	1	–	–	1

The PLA internal logic to implement this circuit is given below

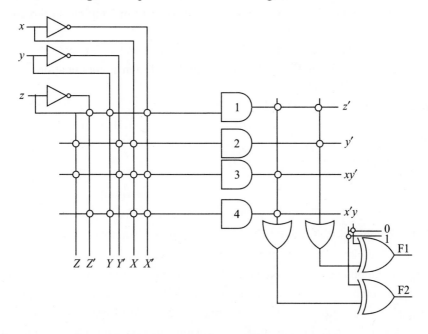

7.16 PROGRAMMABLE ARRAY LOGIC (PAL)

The broad-spectrum structure of PAL is analogous to PLA, but in PAL device only one AND gate is programmable. The OR array in PAL device is predetermined by company. PAL is a category of programmable logic device, which is used to implement Boolean functions in digital logic circuits. The PAL was developed by Monolithic Memories, Inc. (MMI) in March 1978. Programmable Array Logic device is consisted of small PROM core and extra output logic used to implement specific preferred Boolean functions with a small number of components.

EXERCISES

Short Answer Questions

1. What is combinational circuit?
2. What is half adder?
3. Differentiate between half adder and full adder.
4. What is full adder?
5. What is magnitude comparator?
6. What is multiplexer?
7. What is encoder?
8. What is demultiplexer?

9. What is half and full subtractors?
10. Name the combinational circuit which is used to perform addition on three-bit as input and produce two bits output.
11. How many inputs are required in designing seven segment display?
12. How and why inputs and outputs are required to design a combination for converting BCD to Excess-3 code.
13. What is BCD to decimal decoder?
14. How many control signals are used in implementing a Boolean function using 8×1 multiplexer?
15. What is demultiplexer?

Long Answer Questions

1. What is combinational circuit? Explain with suitable example.
2. Design a combination circuit that convert BCD code to Excess-3 code.
3. Design a combinational that converts 4-bit Gray code to binary code.
4. Design a combinational circuit that converts 4-bit binary number to gray code.
5. Design a combinational circuit that convert 8421 code to Excess-3 code.
6. What is multiplexer? Implement the following function with 4×1 multiplexer.
   (i) $f(x, y, z) = \sum(0, 2, 3, 5, 6, 7)$
   (ii) $f(x, y, z) = \sum(2, 3, 6, 7)$
   (iii) $f(y, z, w) = \sum(1, 2, 3, 5)$
   (iv) $f(x, y, w) = \sum(0, 2, 3, 7)$
7. Implement the following function with 4×1 multiplexer using K-map.
   (i) $f(x, y, z) = \sum(0, 2, 4, 5, 6, 7)$
   (ii) $f(x, y, z) = \sum(2, 3, 6, 7)$
   (iii) $f(y, z, w) = \sum(1, 4, 3, 5)$
   (iv) $f(x, y, w) = \sum(0, 2, 3, 4, 7)$
8. Implement the following function with 4×1 multiplexer by comparing the coefficient of control signals
   (i) $f(x, y, x) = x'y + yz + xy' + x'$
   (ii) $f(x, y, z) = y'z + xz + xy' + xy$
   (iii) $f(x, y, z) = xz' + xy + xz + xy'$
   (iv) $f(x, y, z) = yz + xy + x'y + x'y$
   (v) $f(x, y, z) = xz + xy + yz'$

9. Implement the following function with 8×1 multiplexer.
   (i) $f(x, y, z, p) = \sum(0, 2, 3, 5, 6, 7, 11)$
   (ii) $f(x, y, z, w) = \sum(0, 1, 4, 5, 7, 8, 9)$
   (iii) $f(x, y, z, m) = \sum(2, 3, 6, 7, 9)$
   (iv) $f(x, y, z, w) = \sum(1, 2, 3, 5, 11, 12, 13)$
   (v) $f(x, y, A, B) = \sum(0, 2, 3, 11, 12)$
   (vi) $f(x, y, z, w, u) = \sum(1, 2, 3, 5, 11, 12, 13, 16, 18)$
   (vii) $f(x, y, z, w, u, v) = \sum(1, 2, 11, 12, 13, 31, 32)$
10. What is encoder? Design 4×16 decoder.
11. Design a combinational circuit which accepts three-bit input and produces square of its input number.
12. What is full adder? Design a full adder using two half adders.
13. What is full subtractor? Design full adder using two half subtractors.
14. What is magnitude comparator? Design a magnitude comparator to compare two numbers of 5 bits.
15. Design combination circuit 2421 code to BCD code.
16. The control signals of 4×1 multiplexer are y and z, the data variable is x. Draw the logical circuit of 4×1 multiplexer where $I_0 = 1$, $I_1 = 0$, $I_2 = x$ and $I_1 = x'$

Multiple Choice Questions

1. The feedback is required in the following circuit
   (a) Sequential circuit
   (b) Combinational circuit
   (c) (a) and (b)
   (d) None
2. Magnitude comparator is
   (a) Sequential circuit
   (b) Combinational circuit
   (c) Combinational as well as sequential circuit
   (d) None
3. When you are implementing the following Boolean function with 8×1 multiplexer $f(x, y, z, w) = \sum(0, 2, 5, 7, 8, 10, 11, 13, 15)$. The value of I_0, I_6 and I_7 are
   (a) 2, 0, 1
   (b) 0, 1, 1
   (c) 1, w, 1
   (d) w, 1, 0
4. A combinational circuit accepts two inputs and produces two outputs are known as
   (a) Half adder and half subtractor
   (b) Full adder
   (c) Full subtractor
   (d) Decoder

5. Which of the following combinational circuit accept more than one input and produce one output is
   (a) Decoder
   (b) Multiplexer
   (c) Demultiplexer
   (d) Magnitude comparator

6. How many 2 × 4 decoders are used to build 1 × 16 decoder?
   (a) 1
   (b) 2
   (c) 3
   (d) 4

7. The binary numbers $A = 101$ and $B = 110$ are used as inputs of a magnitude comparator. What are the output levels?
   (a) $A = B = 1, A < B = 1, A < B = 0$
   (b) $A > B = 1, A < B = 0, A < B = 1$
   (c) $A < B = 1, A = B = 0, A > B = 1$
   (d) $A = B = 1, A < B = 1, A < B = 1$

8. Find the Boolean expression for logical circuit of the following block diagram of multiplexer.

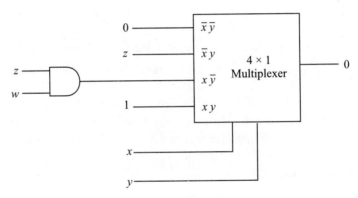

   (a) $f(x, y, x) = x'yz + xy'zw + xy$
   (b) $f(x, y, x) = x'yz + xy' + xy$
   (c) $f(x, y, x) = x'yz + y'zw + xy$
   (d) $f(x, y, x) = x'yz + xzw + xy$

9. A combinational circuit accepts x^n and less than its number of input lines and produces n number of output lines is known as
   (a) Decoder
   (b) Multiplexer
   (c) Encoder
   (d) Half adder

10. Generally demultiplexer circuit facilitates which of the following?
    (a) Decimal to hexadecimal
    (b) Single input multiple output
    (c) Even parity to odd parity
    (d) Odd parity to even parity

11. What is the name of the following logical circuit diagram?

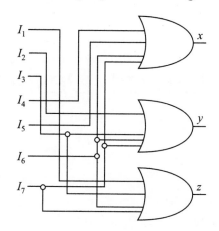

(a) 2 4 decoder
(c) Decimal decoder
(b) 8 3 decoder
(d) 10 4 decoder

Answers

| 1. (b) | 2. (b) | 3. (c) | 4. (a) | 5. (b) | 6. (d) |
| 7. (d) | 8. (a) | 9. (c) | 10. (b) | 11. (b) | |

BIBLIOGRAPHY

http://www.globalspec.com/reference/4276/348308/section-6-5-programmable-array-logic-pal-devices.

Mana, M.M. and Ciletti, M.D., *Digital Design*, 4th ed., Pearson.

CHAPTER 8

Sequential Circuit Design

8.1 INTRODUCTION

A circuit which required feedback is known as sequential circuit. This circuit is used in memory design. The flip flop is example of sequential circuit. One flip flop is capable to store one bit information. In another word you can say the flip flop is connected each other to solve a particular purpose. The block diagram of sequential circuit is given in Figure 8.1.

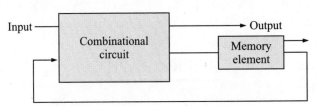

Figure 8.1 Sequential circuit block diagram.

8.2 FLIP FLOPS

Flip flop is a sequential circuit, which is capable to store one bit information. The flip flop is used a component of main memory. The primary memory is designed from flip flops. The block diagram of flip flops is given in Figure 8.2. The flip flop has four lines, Input, Output, Read/Write and Select lines. Input element is used to send data to flip flop. The output is used to get output from flip flop. The read line is used to read the data from flip flop and write line is used to send signal for writing data in flip. The select line is used to select the particular flip flop in which data is given, data is taken or read and write data to flip flop. The one flip flop is also known as cell, which only stores one bit information. The first flip flop was developed by William Eccles and F.W. Jordanin 1918, this circuit was called the Eccles–Jordan trigger circuit.

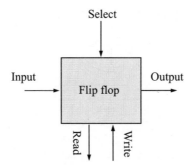

Figure 8.2 Flip flop block diagram.

8.3 TYPES OF FLIP FLOPS

Various types of flip flops are given below.

  (i) RS flip flop
 (ii) Clocked RS flip flop
(iii) JK flip flop
 (iv) D type flip flop
  (v) T type flip flop

(i) RS flip flop

The RS flip flop is the simplest doable memory component. It is created by feeding the outputs of two NOR gates back to the opposite NOR gates input. The inputs R and S are unit remarked for the Reset and Set inputs. To know the operation of the RS flip flop (or RS-latch) take into account the subsequent scenarios as $S = 1$ and $R = 0$: The output of rock bottom NOR gate is adequate to zero, $Q' = 0$. Therefore both the inputs to the highest NOR gate are adequate to one, thus, $Q = 1$. Hence, the input combination $S = 1$ and $R = 0$ ends up in the flip flop being set to $Q = 1$. If $S = 0$ and $R = 1$ then it is just like the arguments given above, the outputs become $Q = 0$ and $Q' = 1$. We are saying that the flip flop is reset. In the case of $S = 0$ and $R = 0$, assume the flip flop is ready ($Q = 0$ and $Q' = 1$), then the output of the highest NOR gate remains at $Q = 1$ and also the lowest NOR gate stays at $Q' = 0$. Similarly, once the flip flop is in a very reset state ($Q = 1$ and $Q' = 0$), it will stay there with this input combination. Therefore, with inputs $S = 0$ and $R = 0$, the flip flop remains in its state. If $S = 1$ and $R = 1$, this input combination should be avoided. The truth table of RS flip flop is given in Table 8.1.

Table 8.1 RS flip flop truth table

R	S	Q	Q'	Comment
0	0	0	0	Hold state
0	1	1	0	Set
1	0	0	1	Reset
1	1	?	?	Avoid, No output

The logical circuit of flip flop is given in Figure 8.3.

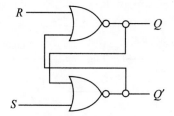

Figure 8.3 Flip flop logical circuit.

The construction RS flip flop with NAND gates is given in Table 8.2.

Table 8.2 RS flip flop with NAND gate truth table

R	S	Q	Q'	Comment
0	0	0	0	Hold state
0	1	1	0	Set
1	0	0	1	Reset
1	1	?	?	Avoid, No output

The circuit diagram of RS flip flop with NAND gate is given in Figure 8.4.

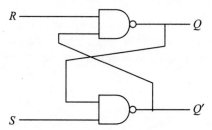

Figure 8.4 RS flip flop with NAND gate circuit diagram.

The block diagram of RS flip flop is given in Figure 8.5.

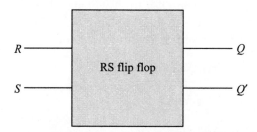

Figure 8.5 RS flip flop block diagram.

(ii) Clocked RS flip flop

The clocked RS flip flop is extension of RS flip flop. The clocked pulse is added with the help of two AND gates in RS flip flop to make clocked RS flip flop. The logical diagram of clocked flip flop is given in Figure 8.6.

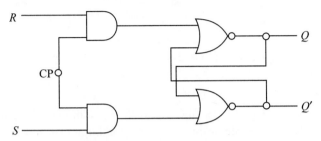

Figure 8.6 Clocked RS flip flop.

Truth table of clocked flip flop is given in Table 8.3.

Table 8.3 Clocked RS flip flop table

Q	R	S	Q(t + 1)	Comment
0	0	0	0	Hold state
0	0	1	1	Set
0	1	0	0	Reset
0	1	1	X	Indeterminate, No output
1	0	0	1	Hold state
1	0	1	1	Set
1	1	0	0	Reset
1	1	1	X	Indeterminate, No output

K-map for $Q(t + 1)$ is given in Figure 8.7.

Q/RS	00	01	11	10
0		1	X	
1	1	1	X	

Figure 8.7 K-map for $Q(t + 1)$.

The equation obtained from K-map is $Q(t + 1) = S + R'Q$.

The **state equation** of clocked flip flop is $Q(t + 1) = S + R'Q$.

The block diagram of clocked flip flop is given in Figure 8.8.

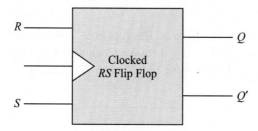

Figure 8.8 Clocked RS flip flop block diagram.

(iii) JK flip flop

The JK flip flop is largely a gated RS flip flop with the addition of a clock input electronic equipment that stops the amerceable or invalid output condition that may occur once both inputs S and R area unit adequate to logic level 1 owing to this extra clocked input, a JK flip flop has four doable input combos, 1, 0, no change and toggle. The logical diagram of JK flip flop is design by replacing R with J, S with K and adding Q to J and Q' to K in RS flip flop. The JK flip flop is developed by Jack Kilby, therefore, the initial letter of hi two word name replaces R and S like $J = S$ and $K = R$. The logical diagram of JK flip flop is given in Figure 8.9.

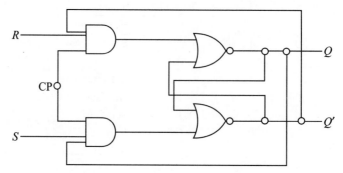

Figure 8.9 JK flip flop logical diagram.

The **truth table** of JK flip flop is given in Table 8.4.

Table 8.4 JK flip flop truth table

Q	J	K	Q(t + 1)	Comment
0	0	0	0	Hold state
0	0	1	0	Set
0	1	0	1	Reset
0	1	1	1	Output
1	0	0	1	Hold state
1	0	1	0	Set
1	1	0	1	Reset
1	1	1	0	Output

K-map for $Q(t + 1)$ is given in Figure 8.10.

QJK	00	01	11	10
0		1	1	1
1	1			1

Figure 8.10 K-map for $Q(t + 1)$.

$$Q(t + 1) = Q'J + QK'$$

The state equation of JK flip flop is $Q(t + 1) = Q'J + QK'$
The block diagram of JK flip flop is given in Figure 8.11.

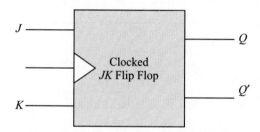

Figure 8.11 JK flip flop block diagram.

(iv) D type flip flop

In this type of flip only one input is used as D. This type of flip flop is designed with the help of JK flip flop by adding J and K together. One NAND gate is used to invert the input place between J and K. The logical diagram of D type flip flop is given in Figure 8.12.

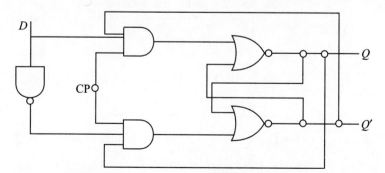

Figure 8.12 D type flip flop logical diagram.

The truth table of D type flip flop is given in Table 8.5. The block diagram of D type flip flop is given in Figure 8.13.

Table 8.5 D type flip flop truth table

Q	D	$Q(t+1)$
0	0	0
0	1	1
1	0	0
1	1	1

The state equation is $Q(t+1) = D$

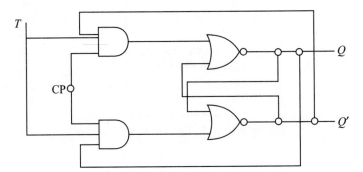

Figure 8.13 D type flip flop block diagram.

(v) T type flip flop

T type flip flop is derived from D type flip flop. In this type of flip flop only one input is used. The addition NAND gate is used to invert the number in D type flip flop and is removed when it became T type flip flop. The logical diagram of T type flip flop is given in Figure 8.14.

Figure 8.14 T type flip flop logical diagram.

The truth table of T type flip flop is given in Table 8.6.

Table 8.6 T type flip flop truth table

Q	T	$Q(t+1)$
0	0	0
0	1	1
1	0	1
1	1	0

The state equation is $Q(t+1) = Q'T + QT'$. The block diagram of T type flip flop is given in Figure 8.15.

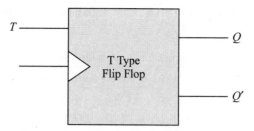

Figure 8.15 *T* type flip flop block diagram.

8.4 SEQUENTIAL CIRCUIT DESIGN PROCEDURE

8.4.1 Design Procedure in Case of State Equation

The following are the steps to design a sequential circuit if state equation is given in the problem.

Step 1: Study the problem carefully and find the all state equations given in problem and assign equation number to each state equation.

Step 2: Find the number flip flop required in designing the sequential circuit, number of flip flop is equal to the number of state equation.

Step 3: If type of flip flop is given in problem for designing the sequential circuit, then write the state equation of that flip flop and assign the equation number in respect to state equation given in the problem. If flip flop is not given, then assume any flip flop as per your choice and write its state equation and give equation number.

Step 4: Compare the coefficient of state equation of a particular flip flop to each state equation given in the problem to find the input of each flip flop.

Step 5: Draw the sequential with the input obtained in Step 4.

EXAMPLE 8.1 Design a sequential circuit using JK flip flop from the following state equations:

$$A(t + 1) = A'BC + ABC' + BC$$
$$B(t + 1) = AC' + B'C + ABC$$
$$C(t + 1) = AC + AB' + BC$$

Solution: There are three state equations, therefore, three flip flops *A*, *B* and *C* are required to design the circuit.

The state equation JK flip flop is

$$Q(t + 1) = JQ' + K'Q \qquad \ldots(1)$$

Replace Q with A in Eq. (1) for getting state equation of flip flop A

$$A(t + 1) = J_A A' + K'_A A \qquad \ldots(2)$$

Replace Q with B in Eq. (1) for getting state equation of flip flop B

$$B(t + 1) = J_B B' + K'_B B \qquad \ldots(3)$$

188 Digital Logic Design

Replace Q with C in Eq. (1) for getting state equation of flip flop C

$$C(t + 1) = J_C C' + K'_C C \qquad \ldots(4)$$

Now convert each of the given equation in standard state equation form of JK flip flop

$$\begin{aligned}
A(t + 1) &= A'BC + ABC' + BC \\
&= A'BC + ABC' + (A + A')BC \\
&= A'BC + ABC' + ABC + A'BC \\
&= A'BC + A'BC + ABC' + ABC \\
&= (BC + BC)A' + (BC' + BC)A \\
&= (BC)A' + (B(C' + C))A \\
&= (BC)A' + (B)A \qquad \ldots(5)
\end{aligned}$$

$$\begin{aligned}
B(t + 1) &= AC' + B'C + ABC \\
&= A(B + B')C' + B'C + ABC \\
&= ABC' + AB'C' + B'C + ABC \\
&= AB'C' + B'C + ABC + ABC' \\
&= (AC' + C)B' + (AC + AC')B \\
&= (AC' + C)B' + (AC + AC')B \\
&= (A + C)B' + (A(C + C'))B \\
&= (A + C)B' + (A)B \qquad \ldots(6)
\end{aligned}$$

$$\begin{aligned}
C(t + 1) &= AC + AB' + BC \\
&= AC + AB'(C + C') + BC \\
&= AB'C' + BC + AC + AB'C \\
&= AB'C' + (B + A + AB')C \\
&= AB'C' + (B + A(1 + B'))C \\
&= AB'C' + (B + A)C \qquad \ldots(7)
\end{aligned}$$

Now equating Eqs. (2) and (5), we get

$$J_A A' + K'_A A = (BC)A' + (B)A$$

By comparing the coefficients of A and A', we get

$$J_A = BC, K'_A = B, K_A = B'$$

Equating Eqs. (3) and (6), we get

$$J_B B' + K'_B B = (A + C)B' + (A)B$$

Comparing coefficients of both sides, we get

$$J_B = A + C \text{ and } K'_B = A, K_B = A'$$

Equating Eqs. (4) and (7), we get

$$J_C C' + K'_C C = AB'C' + (B + A)C$$

Comparing coefficients of both sides, we get
$$J_C = AB' \text{ and } K_C' = (B + A), K_C = (B + A)'$$
The inputs of flip flops A, B and C are
$$J_A = BC \text{ and } K_A = B'$$
$$J_B = A + C \text{ and } K_B = A'$$
$$J_C = AB' \text{ and } K_C = (B + A)'$$
Now draw the sequential circuit.

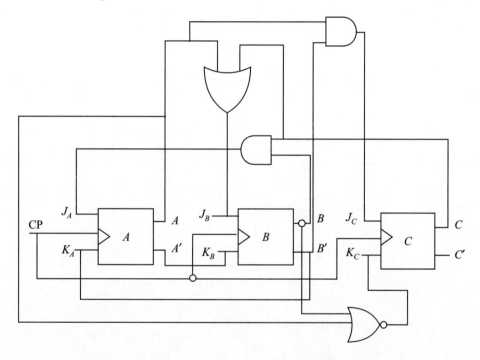

EXAMPLE 8.2 Design a sequential circuit with four flip flops X, Y, Z and W. The next states of X and Y are $Z + W$ and $Z.W$, respectively, and the next states of Z and W are Y and $Y + X$, respectively.

Solution: The state equation obtained from question is as follows:
$$X(t + 1) = Z + W \qquad \qquad ...(1)$$
$$Y(t + 1) = Z.W \qquad \qquad ...(2)$$
$$Z(t + 1) = Y \qquad \qquad ...(3)$$
$$W(t + 1) = Y + X \qquad \qquad ...(4)$$

The name of flip flop is not mentioned in the question, so we can choose D type flip flop, because it is simple. The state equation of D type flip flop is

$Q(t + 1) = D$, Replace Q with X, Y, Z and W to get the following equation:

$$X(t + 1) = D_X \qquad \text{...(5)}$$
$$Y(t + 1) = D_Y \qquad \text{...(6)}$$
$$Z(t + 1) = D_Z \qquad \text{...(7)}$$
$$W(t + 1) = D_W \qquad \text{...(8)}$$

Equating Eqs. (1)–(4) with Eqs. (5)–(8), we get

$$D_X = Z + W$$
$$D_Y = Z.W$$
$$D_Z = Y$$
$$D_W = Y + X$$

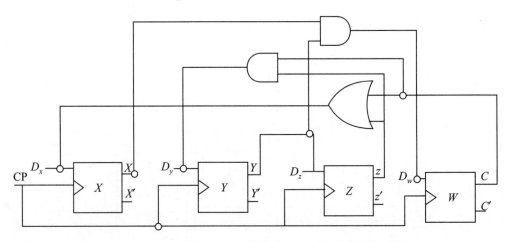

8.4.2 Design Procedure in Case of State Table

If the state table is given in question, then you can follow this procedure.

Step 1: Find the number of flip flop required to design sequential circuit from state table. The number of bit in present state is equal to the number flip flop required to design the sequential circuit.

Step 2: Write the excitation table from the state table and the excitation table of particular flip flop.

Step 3: Use K-map to find the minimized flip flop of each input.

Step 4: Draw sequential circuit with the equation obtained in Step 3.

The excitation tables of different types of flip flops are given in Tables 8.7–8.10.

Table 8.7 JK flip flop excitation table

Q	Q(t+1)	R	S
0	0	X	0
0	1	0	1
1	0	1	0
1	1	0	X

Table 8.8 RS flip flop excitation table

Q	Q(t+1)	J	K
0	0	0	X
0	1	1	X
1	0	X	1
1	1	X	0

Table 8.9 D type flip flop excitation table

Q	Q(t+1)	D
0	0	0
0	1	1
1	0	0
1	1	1

Table 8.10 T type flip flop excitation table

Q	Q(t+1)	T
0	0	0
0	1	1
1	0	1
1	1	0

EXAMPLE 8.3 Design a sequential circuit with the following state table. Use RS flip flop.

Present State	Next State
AB	AB
00	10
01	11
10	01
11	00

Solution: There are two bits available in present state, therefore, two flip flops are required. The excitation table is given below.

Present State	Next State	Flip Flop Input			
AB	AB	R_A	S_A	R_B	S_B
00	10	0	1	X	0
01	11	0	1	0	X
10	01	1	0	0	1
11	00	1	0	1	0

$$R_A = AB' + AB = A(B' + B) = A$$
$$S_A = A'B' + A'B = A'(B' + B) = A'$$
$$R_B = A'B' + AB$$
$$S_B = AB' + AB' = AB'$$

192 Digital Logic Design

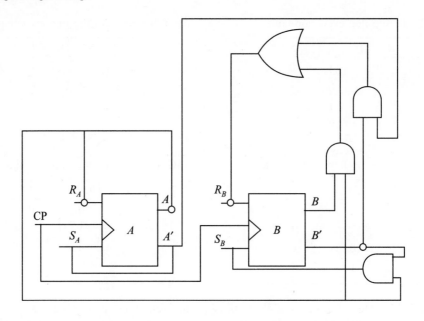

EXAMPLE 8.4 Design a sequential circuit with the following state table. Use T flip flop.

Present State AB	Next State	
	$X=0$ AB	$X=1$ AB
00	10	01
01	11	00
10	01	11
11	00	10

Solution: There are two bits available in present state, therefore, two flip flops are required. The excitation table is given below.

Present State			Next State			
A	B	X	A	B	T_A	T_B
0	0	0	1	0	1	1
0	0	1	0	1	0	1
0	1	0	1	1	1	0
0	1	1	0	0	0	1
1	0	0	0	1	1	1
1	0	1	1	1	0	1
1	1	0	0	0	1	1
1	1	1	1	0	0	1

K-map for T_A

A/BX	00	01	11	10
0	1			1
1	1			1

$T_A = X'$

K-map for T_B

A/BX	00	01	11	10
0	1	1	1	
1	1	1	1	1

$T_B = B' + X + AB$

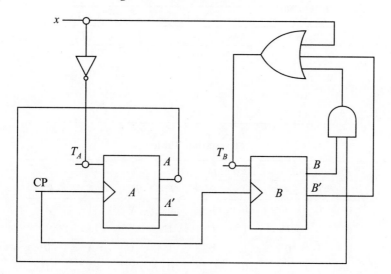

8.4.3 Design Procedure in Case of State Diagram

If the state diagram is given in problem, then following steps are followed.

Step 1: Draw state table from given state diagram.

Step 2: Draw excitation table from state table.
Step 3: Use K-map to find the minimized input equation of flip flop.
Step 4: Draw the sequential circuit for the equation obtained in Step 3.

EXAMPLE 8.5 Design a sequential circuit with the following state diagram.

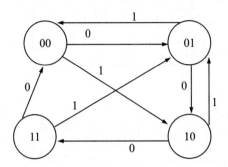

Solution: Each state having two bits, therefore two flip flops are required to design a sequential circuit. Now convert this state diagram into state table. Use any type of flip flop, because the name of flip flop is not mentioned in the question. Use JK flip flop.

State Table

Present State	Next State	
	X = 0	X = 1
AB	AB	AB
00	01	11
01	11	00
10	00	01
11	10	01

Excitation Table

Present State			Next State		J_A	K_A	J_B	K_B
A	B	X	A	B				
0	0	0	0	1	0	X	1	X
0	0	1	1	1	1	X	1	X
0	1	0	1	1	1	X	X	0
0	1	1	0	0	0	X	X	1
1	0	0	0	0	X	1	0	X
1	0	1	0	1	X	1	1	X
1	1	0	1	0	X	0	X	1
1	1	1	0	1	X	1	X	0

K-map for J_A

AB/x	0	1
00		1
01	1	
11	x	x
10	x	x

$$J_A = BX' + B'X$$

K-map for K_A

AB/x	0	1
00	x	x
01	x	x
11		1
10	1	1

$$K_A = X + B'X$$

K-map for J_B

AB/x	0	1
00	1	1
01	x	x
11	x	x
10		1

$$J_B = A' + X$$

196 Digital Logic Design

K-map for K_B

AB/x	0	1
00	x	x
01		1
11	1	
10	x	x

$K_B = A'X + AX'$

$J_A = BX'$
$K_A = X + B'X$
$J_B = A' + X$
$K_B = A'X + AX'$

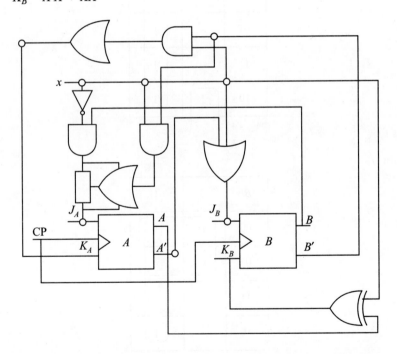

EXAMPLE 8.6 Design a sequential circuit with the following state diagram. Use D type flip flop.

Sequential Circuit Design

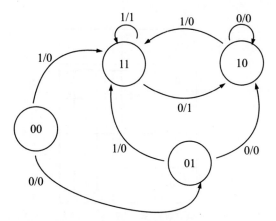

Solution: Each state having two bits, therefore two flip flops are required to design sequential circuit. Now convert this state diagram into state table.

Present State AB	Input X	Next State AB	Output Y
00	0	01	0
00	1	11	0
01	0	10	0
01	1	11	0
10	0	10	0
10	1	11	0
11	0	10	1
11	1	11	1

K-map for D_A

AB/x	0	1
00		1
01		1
11	1	1
10	1	1

$$D_A = A + x$$

K-map for D_B

AB/x	0	1
00	1	1
01		1
11		1
10		1

$D_A = A'B' + x$

The logical circuit is given below:

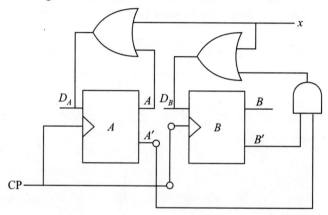

EXAMPLE 8.7 Draw the state table and state diagram for the following sequential circuit.

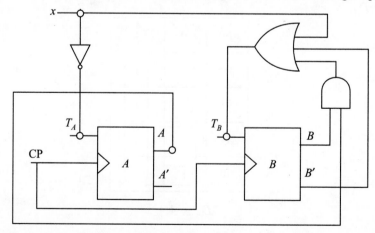

Solution: There are two flip flops used in the diagram, so two bits are available in present state and next state. The X is input in the circuit.

$$T_A = X'$$
$$T_B = B' + X + AB$$

The excitation table of the above circuit is obtained from $T_A = X'$, $T_B = B' + X + AB$ and the excitation table of T flip flop.

Q	Q(t+1)	T
0	0	0
0	1	1
1	0	1
1	1	0

The excitation table of the sequential circuit is as follows:

Present State			Next State			
A	B	X	A	B	T_A	T_B
0	0	0	1	0	1	1
0	0	1	0	1	0	1
0	1	0	1	1	1	0
0	1	1	0	0	0	1
1	0	0	0	1	1	1
1	0	1	1	1	0	1
1	1	0	0	0	1	1
1	1	1	1	0	0	1

The state table is derived from the above excitation table.

Present State		Next State X=0		Next State X=1	
A	B	A	B	A	B
0	0	1	0	0	1
0	1	1	1	0	0
1	0	0	1	1	1
1	1	0	0	1	0

The state diagram is as follows:

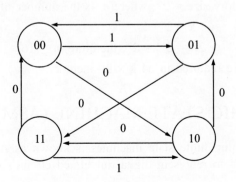

8.5 RANDOM ACCESS MEMORY (RAM)

The RAM is volatile primary memory. The contents of RAM will be destroyed if light is switched off. One flip flop is used to store one bit information in RAM. The flip flop is used to design Random Access Memory (RAM). Many flip flops are connected in such a way that data can be accessed randomly. The memory stored the group of binary information is called word. Let us consider a 4 × 8 RAM. In this RAM 32 flip flops are required. Four register of 8 bits are connected to give a 4 × 8 RAM. The block diagram of 4 × 8 is RAM given in Figure 8.17. The block diagram of one cell is given in Figure 8.16. RAM is a semiconductor memory. In RAM data is read or written randomly, it means data is accessed randomly. A memory is a group of 1 and 0 used to represent data, instruction and alphanumeric. The group of 8 bits is known as bytes. The capacity of RAM is calculated in the form of bytes. The 16 bits mean 2 bytes.

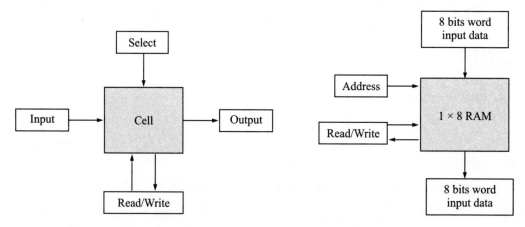

Figure 8.16 Cell block diagram. **Figure 8.17** 4 × 8 RAM.

In this example total memory size is 32 bits. The address is determined with the formula memory size = 2^m, where m is address line. The memory size of this RAM is 32 k, so $32 = 2^m, 2^5 = 2^m$. Therefore address $m = 5$.

EXAMPLE 8.8 Find the number of address line for the memory 64 k × 20.

Solution: We know memory size = 2^m, where m is the number of address line.

So, $64 \text{ k} = 2^m$

$2^6 \times 2^{10} = 2^{16} = 2^m$, therefore $m = 16$.

So, 16 address lines are required in 64 k × 20 memory.

8.6 ALGORITHMIC STATE MACHINE (ASM)

A method used for designing finite state machines is known as Algorithmic State Machine (ASMs). The ASM is used to represent digital integrated circuit's diagram. The Algorithmic

State Machine diagram is similar to state diagram but less prescribed and as a result easier to understand. Describing the sequential operations of a digital system, an ASM chart method is used. The following are the various symbols used in ASM chart.

State Box: It is in a rectangular shape, in which one input and one output are there. This box is used for one or more operations. The symbol of state box is given in Figure 8.18.

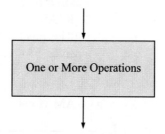

Figure 8.18 State box symbol.

Decision Box: It is represented by diamond shape. Decision box has one input and many outputs. It is used to enumerate multiple paths, which can be followed. The symbol of decision box is given in Figure 8.19.

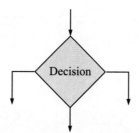

Figure 8.19 Decision box.

Conditional Box: It is represented by rectangular shape with rounded corner. Decision Box is always followed by conditional box. It contains one or multiple conditional operations. The symbol of conditional box is given in Figure 8.20.

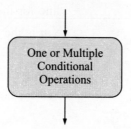

Figure 8.20 Conditional box.

202 Digital Logic Design

EXAMPLE 8.9 Draw ASM chart from the following state table

Present State AB	Next State X = 0 AB	Next State X = 1 AB
00	10	01
01	11	00
10	01	11
11	00	10

Solution: The ASM chart is given below.

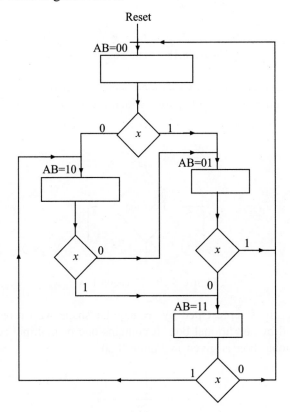

EXERCISES

Short Answer Questions

1. What is flip flop?
2. What are the various types of flip flops?
3. What is sequential circuit?

4. List the name of various flip flops.
5. What is the difference between RS and Clocked RS flip flops?
6. What is the difference between D type and T type flip flops?
7. What do you mean by state table?
8. What do you mean by state diagram?
9. In JK flip flop, J and K stand for what?
10. In RS flip flop, R and S stand for what?
11. What is latch?

Long Answer Questions

1. Explain Clocked RS flip flop.
2. What is state equation of JK, RS, D type and T type flip flops?
3. Explain excitation table of JK, RS, T type and D type.
4. Design sequential circuit with four flip flops A, B, C and D. The next state of B and C are A and D, respectively. The next state of A is X-NOR of B and C. The next state of D is BC.
5. Prove that the state equation of T type is $Q(t + 1) = TQ' + T'Q$.
6. Prove that the state equation of JK flip flop is $Q(t + 1) = JQ' + K'Q$.
7. Covert the following state table into state diagram.

Present State	Next State	
	$X = 0$	$X = 1$
A	B	C
B	A	D
C	B	A
D	C	B

8. Convert the following state diagram into state table.

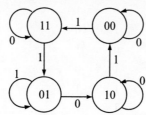

9. Design a sequential circuit from the following state diagram given below using T type flip flop

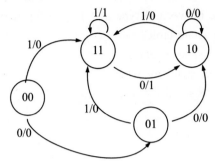

10. Design a sequential circuit from the following state equations using JK flip flop.

$$A(t + 1) = B'C + AC + ABC$$
$$B(t + 1) = AB + BC$$
$$C(t + 1) = A + B$$

11. Design sequential circuit from the state table given below using RS flip flop.

Present State	Next State	
	$X = 0$	$X = 1$
000	001	010
001	100	110
010	101	011
110	011	000

12. The excitation table is given in the following table. The value of R_A, S_A, R_B and S_B are missing from the table. Fill the value of R_A, S_A, R_B and S_B in the following table.

Present			Next State		Input of Flip Flop			
A	B	Y	A	B	R_A	S_A	R_B	S_B
0	0	0	0	0				
0	0	1	0	1				
0	1	0	1	0				
0	1	1	0	1				
1	0	0	1	0				
1	0	1	1	1				
1	1	0	1	1				
1	1	1	0	0				

13. Prove that the state equation of RS flip flop is $Q(t + 1) = S + R'Q$.
14. Prove that the state equation of D type flip flop is $Q(t + 1) = D$.

15. Find state equation, state diagram and state table of the following sequential circuit.

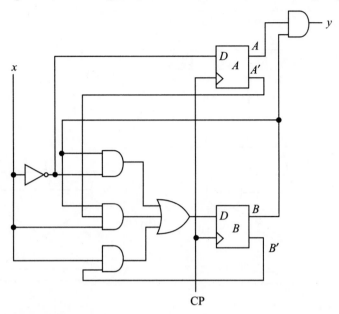

Multiple Choice Questions

1. Flip flop is a
   - (a) Combinational circuit
   - (b) Sequential circuit
   - (c) Magnitude comparator circuit
   - (d) None

2. Which of the following state equation of RS flip flop?
   - (a) $Q(t + 1) = S + R'Q$
   - (b) $Q(t + 1) = S + RQ'$
   - (c) $Q(t + 1) = S + RQ$
   - (d) $Q(t + 1) = S' + RQ$

3. The state equation $Q(t + 1) = QK' + Q'J$ of
   - (a) RS flip flop
   - (b) JK flip flop
   - (c) T type flip flop
   - (d) D type flip flop

4. How many flip flops are required to design a sequential circuit using the following state equations:
   $$A(t + 1) = A + AB' + A'C$$
   $$B(t + 1) = AB'C + BC$$
   $$C(t + 1) = AC + AB'$$
   - (a) 3
   - (b) 2
   - (c) 4
   - (d) 5

5. How many flip flops are required to design an 8 × 32 RAM?
   - (a) 257
   - (b) 256
   - (c) 255
   - (d) 16

6. The state diagram of the following state table is

Present State		Next State X = 0		Next State X = 1	
A	B	A	B	A	B
0	0	1	0	0	1
0	1	1	1	0	0
1	0	0	1	1	1
1	1	0	0	1	0

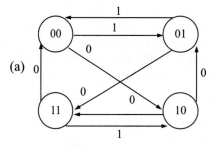
(a)

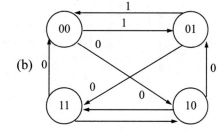
(b)

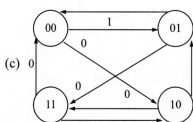
(c)

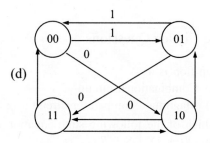
(d)

7. How many flip flops are required to design a sequential circuit from the state table given in Question 6?
(a) 3 (b) 4
(c) 2 (d) 1

8. How many self loops in the following state diagram?

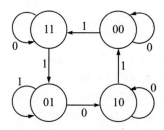

(a) 2 (b) 3
(c) 4 (d) 5

9. The sequential circuit is designed with state equation $A(t + 1) = A + AB' + A'C$. What is the value of J_A and K_A?
   (a) $J_A = 1 + B'$ and $K_A = C$
   (b) $J_A = B'$ and $K_A = C'$
   (c) $J_A = C$ and $K_A = (1 + B')'$
   (d) $J_A = 1 + B'$ and $K_A = B + C$

10. If $J_A = 1 + B'$ and $K_A = C$, the state equations is:
    (a) $A(t + 1) = A + AB' + A'C$
    (b) $A(t + 1) = A' + A'B' + AC'$
    (c) $A(t + 1) = A + B' + A'C$
    (d) $A(t + 1) = A + AB' + C$

11. If $D_A = BC + B$, then the state equation is:
    (a) $A(t + 1) = B' + A'C$
    (b) $A(t + 1) = A'B' + AC'$
    (c) $A(t + 1) = A + B' + A'C$
    (d) $A(t + 1) = BC + C'$

12. The value of T_A and T_B from the following sequential circuit is:

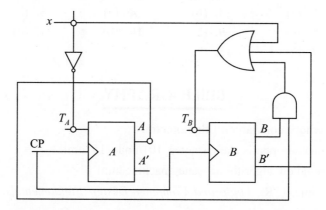

    (a) $T_A = X'$ and $T_B = AB + B' + X$
    (b) $T_A = X$ and $T_B = AB + B' + X$
    (c) $T_A = X'$ and $T_B = AB + B$
    (d) $T_A = X'$ and $T_B = AB + B + X$

13. The value of J_B from the following sequential circuit is:

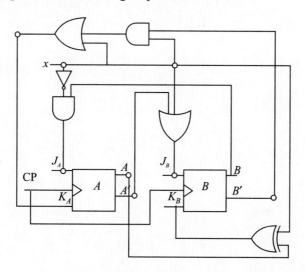

(a) $J_B = A + X$ (b) $J_B = A X$
(c) $J_B = A + X$ (d) $J_B = A + XA$

14. What is the value of K_A in the sequential circuit shown in Question 13.
    (a) $K_A = B X + X$ (b) $K_A = B + X$
    (c) $K_A = BX + X$ (d) $K_A = A + X$

15. The sequential circuit given in Question 13 is
    (a) Asynchronous sequential circuit
    (b) Synchronous sequential circuit
    (c) Asynchronous sequential circuit as well as synchronous sequential circuit
    (d) None

Answers

1. (b)	2. (a)	3. (b)	4. (a)	5. (b)	6. (a)
7. (c)	8. (c)	9. (c)	10. (b)	11. (d)	12. (a)
13. (c)	14. (a)	15. (b)			

BIBLIOGRAPHY

http://www.computerhope.com/jargon/f/flipflop.htm.

http://thalia.spec.gmu.edu/~pparis/classes/notes_101/node115.html.

http://barrywatson.se/dd/dd_algorithmic_state_machine.html

Mano, M.M. and Kime, C.R., *Logic and Computer Design Fundamentals*, 2nd ed., Prentice Hall, 2000.

CHAPTER 9

Counter Design

9.1 COUNTER

A counter is a sequential circuit used for counting the number of clock pulses. The counter is also used to measuring the frequency as well as time period. The counter is a specific type of sequential circuit. There are broadly two types of counter available in real world. The first one is Synchronous, it is also known as parallel, and second one is Asynchronous, it is also known as Ripple.

- **Synchronous or Parallel Counter:** A sequential circuit is known as synchronous or parallel counter, if common clock pulse is given to all flip flops in the circuit.
- **Asynchronous or Ripple Counter:** A sequential circuit is known as asynchronous or ripple counter, if common clock pulse is not given to all flip flops in the circuit.

9.1.1 Applications of Counter

The counter is used for the various purposes in real world. The following are the applications of counter.

- The counter is used as simple clocks to keep track of timing.
- The counter is used to record how many times something has happened, means how many bits have been sent or received while communicating the message and how many steps have been used in some computation.
- The counter is used in processors to keep the track of instruction execution that counter is known as PC (Program Counter). Program Counter consists of a list of instructions that are to be executed one after another instruction. The Program Counter also keeps track of the instruction currently being executed and increments by one on every clock cycle for the next program instruction is then executed.
- The counter is also used to count the different type of sequences.

9.2 BINARY COUNTER

The binary counter is used to count the binary sequence. In the binary counter the output value increases by one on every clock pulse or clock signal. After getting the largest value in that sequence, the output comes back to zero from where the sequence starts. Consider an example; assume a binary sequence of two bits, the present state of this sequence is 00, 01, 10 and 11 and next state of this sequence is 10, 11, 01, 00. This sequence is shown in tabular form as stable table and state diagram are given in Table 9.1 and Figure 9.1.

Table 9.1 State Table

Present State		Next State	
A	B	A	B
0	0	1	0
0	1	1	1
1	0	0	1
1	1	0	0

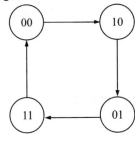

Figure 9.1 State diagram.

The state equation obtained from above state table is as follows.

$$A(t + 1) = A'B' + A'B = A'(B' + B) = A'$$
$$B(t + 1) = A'B + AB'$$

The counter circuit according to the above state equation is given in Figure 9.2.

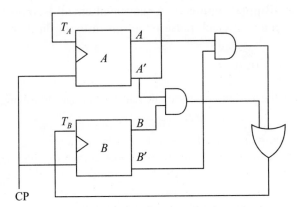

Figure 9.2 Counter circuit.

9.3 SYNCHRONOUS UP/DOWN COUNTER

Consider the three bits in Up/Down counter. In this counter three flip flops are required to accommodate three bits. In this counter, assume a variable x, if the value of x is zero then it will count up and in case of value of x is one (1) then it will count down. The state diagram of Up/Down counter is given in Figure 9.3. The stable table of Up/Down counter is given in Table 9.2.

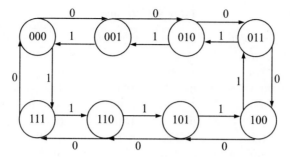

Figure 9.3 Up/Down counter state diagram.

Table 9.2 Up/Down counter stable table

Present State ABC	X	Next Present ABC
000	0	001
000	1	111
001	0	010
001	1	000
010	0	011
010	1	001
011	0	100
011	1	010
100	0	101
100	1	011
101	0	110
101	1	100
110	0	111
110	1	101
111	0	000
111	1	110

Use any type flip to design the Up/Down counter from the above state table. But we assume T type flip flops to design the Up/Down counter. The excitation table from the above state table is given in Table 9.3.

Table 9.3 Excitation table of state Table 9.2

Present State ABC	X	Next Present ABC	T_A	T_B	T_C
000	0	001	0	0	1
000	1	111	1	1	1
001	0	010	0	1	1
001	1	000	0	0	1
010	0	011	0	0	1

(*Contd.*)

Table 9.3 (*Contd.*)

Present State ABC	X	Next Present ABC	T_A	T_B	T_C
010	1	001	0	1	1
011	0	100	1	1	1
011	1	010	0	0	1
100	0	101	0	0	1
100	1	011	1	1	1
101	0	110	0	1	1
101	1	100	0	0	1
110	0	111	0	0	1
110	1	101	0	1	1
111	0	000	1	1	1
111	1	110	0	0	1

K-map for T_A is given in Figure 9.4.

AB/CX	00	01	11	10
00		1		
01				1
11	1	1	1	
10	1		1	1

Figure 9.4 K-map for T_A.

$$T_A = A'B'C'X + A'BCX' + ABC' + AC'X' + ACX + AB'C$$

K-map for T_B is given in Figure 9.5.

AB/CX	00	01	11	10
00		1		1
01		1		1
11		1		1
10		1		1

$$T_B = C'X + CX'$$

Figure 9.5 K-map for T_B.

K-map for T_C is given in Figure 9.6.

AB/CX	00	01	11	10
00	1	1	1	1
01	1	1	1	1
11	1	1	1	1
10	1	1	1	1

Figure 9.6 K-map for T_C.

$T_C = 1$ The logical circuit for synchronous Up/Down counter is given in Figure 9.7.

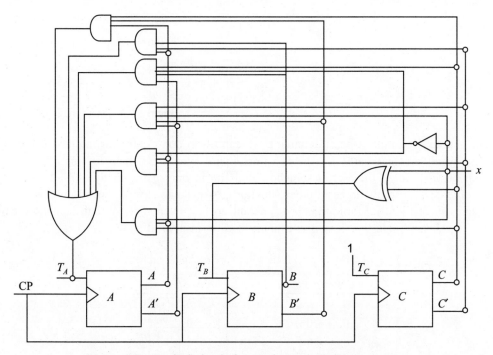

Figure 9.7 Logical circuit for synchronous Up/Down counter.

9.4 UNUSED STATES

In counter designing for a particular sequence in which some states are not used. When you are going to design a counter, in some cases some state will not be used. Consider an example, here is a state table and diagram for a counter that repeatedly counts from 0(000) to 6(110). In this sequence there are three bits which are involved to represent the count from 0 to 6.

But using three bits, the total possible combination will be 7(111). But in this sequence we count upto 6(110) only, so 7(111) is not included in counting. Therefore, the 7(111) will be the unused state. What should we put in the table for this 7(111) state unused states? We can put don't care condition in Table 9.4 in place of unused state to find the minimized equation. The state diagram of this example is given in Figure 9.8. In this state diagram the 111 is not used, so we can say 111 is unused state.

The state table is given in Table 9.4.

Table 9.4 Unfilled don't care in state table

Present state			Next State		
A	B	C	A	B	C
0	0	0	0	0	1
0	0	1	0	1	0
0	1	0	0	1	1
0	1	1	1	0	0
1	0	0	1	0	1
1	0	1	1	1	0
1	1	0	0	0	0
1	1	1	?	?	?

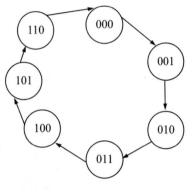

Figure 9.8 State diagram.

For getting the simplest possible circuit, we can fill in don't cares for the next states. This will also result in don't cares for the flip flop inputs, which can help to simplify the input equation of flip flop. If the circuit somehow ends up in one of the many unused states (111 or any other state also available in other situation), its behaviour will depend on exactly what don't cares were filled in Table 9.5. It is guaranteed that even if the circuit somehow enters an unused state in Table 9.5, it will eventually end up in a valid state

Table 9.5 Filled don't care in stable table

Present State			Next State		
A	B	C	A	B	C
0	0	0	0	0	1
0	0	1	0	1	0
0	1	0	0	1	1
0	1	1	1	0	0
1	0	0	1	0	1
1	0	1	1	1	0
1	1	0	0	0	0
1	1	1	X	X	X

9.5 SYNCHRONOUS BCD COUNTER

The synchronous binary counter counts four-bit number from 0000 to 1001. The four any types of flip flops are required to design the binary counter. Let here we use T-type flip flop to design the synchronous binary counter. The state table of binary counter is given in Table 9.6.

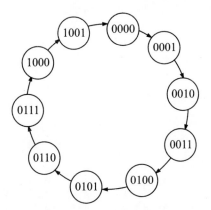

Figure 9.9 Binary counter stable table.

The state table of synchronous binary counter is given in Table 9.6.

Table 9.6 Synchronous binary counter state table

Present State				Next State			
A	B	C	D	A	B	C	D
0	0	0	0	0	0	0	1
0	0	0	1	0	0	1	0
0	0	1	0	0	0	1	1
0	0	1	1	0	1	0	0
0	1	0	0	0	1	0	1
0	1	0	1	0	1	1	0
0	1	1	0	0	1	1	1
0	1	1	1	1	0	0	0
1	0	0	0	1	0	0	1
1	0	0	1	0	0	0	0
1	0	1	0	X	X	X	X
1	0	1	1	X	X	X	X
1	1	0	0	X	X	X	X
1	1	0	1	X	X	X	X
1	1	1	0	X	X	X	X
1	1	1	1	X	X	X	X

Excitation table for the above state table is given in Table 9.7.

Table 9.7 Excitation table of Table 9.6

A	B	C	D	A	B	C	D	T_A	T_B	T_C	T_D
0	0	0	0	0	0	0	1	0	0	0	1
0	0	0	1	0	0	1	0	0	0	1	1
0	0	1	0	0	0	1	1	0	0	0	1
0	0	1	1	0	1	0	0	0	1	1	1
0	1	0	0	0	1	0	1	0	0	0	1
0	1	0	1	0	1	1	0	0	0	1	1
0	1	1	0	0	1	1	1	0	0	0	1
0	1	1	1	1	0	0	0	1	1	1	1
1	0	0	0	1	0	0	1	0	0	0	1
1	0	0	1	0	0	0	0	1	0	0	1
1	0	1	0	X	X	X	X	X	X	X	X
1	0	1	1	X	X	X	X	X	X	X	X
1	1	0	0	X	X	X	X	X	X	X	X
1	1	0	1	X	X	X	X	X	X	X	X
1	1	1	0	X	X	X	X	X	X	X	X
1	1	1	1	X	X	X	X	X	X	X	X

K-map for T_A is given in Figure 9.10.

AB/CD	00	01	11	10
00				
01			1	
11	x	x	x	x
10		1	x	x

Figure 9.10 K-map for T_A.

$$T_A = BCD + AD$$

K-map for T_B is given in Figure 9.11.

AB/CD	00	01	11	10
00			1	
01			1	
11	x	x	x	x
10			x	x

Figure 9.11 K-map for T_B.

$$T_B = CD$$

K-map for T_C is given in Figure 9.12.

AB/CD	00	01	11	10
00		1	1	
01		1	1	
11	x	x	x	x
10			x	x

Figure 9.12 K-map for T_C.

$$T_C = A'D + CD$$

K-map for T_D is given in Figure 9.13.

AB/CD	00	01	11	10
00	1	1	1	1
01	1	1	1	1
11	x	x	x	x
10	1	1	x	x

Figure 9.13 K-map for T_D.

$$T_D = 1$$

The logical diagram of synchronous/parallel BCD counter is given in Figure 9.14.

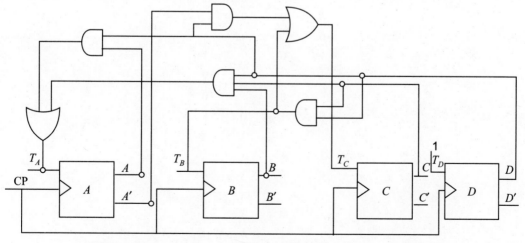

Figure 9.14 Synchronous/parallel *BCD* counter logical diagram.

9.6 DECADE COUNTER

Decade counter is the one which matches through ten distinctive combos of outputs and then resets because the clock takings. We have a tendency to use some type of a feedback in a very 4-bit binary counter to skip any six of the sixteen attainable output states from 0000 to 1111 to urge to a decade counter. A decade counter counts from 0000(0) to 1001(9), means total count ten, it might count as 0000, 0001, 0010, 1000, 1001, 1010, 1011, 1110, 1111, 0000, 0001 so on.

The circuit of decade counter is basically a ripple counter that counts up to sixteen. We have a tendency to want, but a circuit operation within which the count advances from zero to nine and then resets to zero for a replacement cycle. This reset could be accomplished at the required count as follows.

1. With counter REST count = 0000 the counter is prepared to stage counter cycle.
2. Input pulses advance counter in binary sequence up to count of a (count = 1001).
3. Future count pulse advances the count to ten count = 1010. A logic NAND circuit decodes the count of ten providing a level modification at that point to trigger the ammunition unit that then resets all counter stages. Thus, the heartbeat once the counter is at count = nine, effectively leads to the counter about to count = zero.

The four flip flops required to design decade counter because four bits are involved count the sequence for 0000(0) to 1001(9). Let D type flip flop to design decade counter. The excitation table for decade counter is given in Table 9.8.

Table 9.8 Decade counter excitation table

A	B	C	D	D_A	D_B	D_C	D_D
0	0	0	0	0	0	0	1
0	0	0	1	0	0	1	0
0	0	1	0	0	0	1	1
0	0	1	1	0	1	0	0
0	1	0	0	0	1	0	1
0	1	0	1	0	1	1	0
0	1	1	0	0	1	1	1
0	1	1	1	1	0	0	0
1	0	0	0	1	0	0	1
1	0	0	1	0	0	0	0
1	0	1	0	X	X	X	X
1	0	1	1	X	X	X	X
1	1	0	0	X	X	X	X
1	1	0	1	X	X	X	X
1	1	1	0	X	X	X	X
1	1	1	1	X	X	X	X

K-map for D_A is given in Figure 9.15.

AB/CD	00	01	11	10
00				
01			1	
11	x	x	x	x
10	1		x	x

Figure 9.15 K-map for D_A.

$$D_A = BCD + AD'$$

K-map for D_B is given in Figure 9.16.

AB/CD	00	01	11	10
00			1	
01	1	1		1
11	x	x	x	x
10			x	x

Figure 9.16 K-map for D_B.

$$D_B = BC' + BD' + B'CD$$

K-map for D_C is given in Figure 9.17.

AB/CD	00	01	11	10
00		1		1
01		1		1
11	x	x	x	x
10			x	x

Figure 9.17 K-map for D_C.

$$D_C = A'C'D + CD$$

K-map for D_D is given in Figure 9.18.

AB/CD	00	01	11	10
00	1			1
01	1			1
11	x	x	x	x
10	1		x	x

Figure 9.18 K-map for D_D.

The logical circuit diagram of decade counter is given in Figure 9.19.

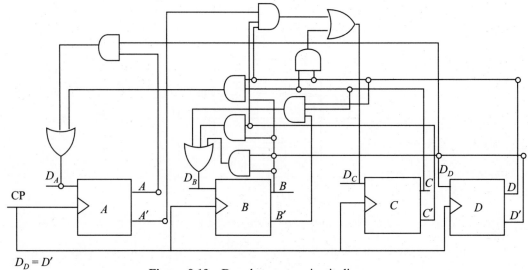

$D_D = D'$

Figure 9.19 Decade counter circuit diagram.

9.7 RING COUNTER

The ring counter is that the simplest examples of a shift register. The best counter is named a hoop counter. The ring counter contains only {1} logical 1 or 0 that it circulates. The entire cycle length is up to the quantity of stages. The ring counter is helpful in applications wherever count must be recognized so as to perform another logic operation. Since only {1} output is ever at logic 1 at given time further logic gates do not seem to be needed to decrypt the counts and also the flip flop outputs are also used on to perform the specified operation. In the ring counter the output of the last flip flop is connected with the input of first flip flop.

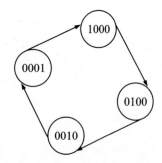

Figure 9.20 Ring counter state diagram.

The arrangement of flip flop in ring counter is in a cascade form. Let design the four stage ring counter, there are four flip flops required. Assume four T type flip flops to design the ring counter are A, B, C and D. The state diagram of ring counter is given in Figure 9.20.

The state table is given in Table 9.9.

Table 9.9 Ring counter state table

	Present State				Next State		
A	B	C	D	A	B	C	D
1	0	0	0	0	1	0	0
0	1	0	0	0	0	1	0
0	0	1	0	0	0	0	1
0	0	0	1	1	0	0	0
0	1	0	0	?	?	?	?
0	1	0	1	?	?	?	?
0	1	1	0	?	?	?	?
0	1	1	1	?	?	?	?
1	0	0	0	?	?	?	?
1	0	0	1	?	?	?	?
1	0	1	0	?	?	?	?
1	0	1	1	?	?	?	?
1	1	0	0	?	?	?	?
1	1	0	1	?	?	?	?
1	1	1	0	?	?	?	?
1	1	1	1	?	?	?	?

Excitation table of ring counter is given in Table 9.10.

Table 9.10 Ring counter excitation table

A	B	C	D	T_A	T_B	T_C	T_D
1	0	0	0	0	1	0	0
0	1	0	0	0	0	1	0
0	0	1	0	0	0	0	1
0	0	0	1	1	0	0	0
0	1	0	0	X	X	X	X
0	1	0	1	X	X	X	X
0	1	1	0	X	X	X	X
0	1	1	1	X	X	X	X
1	0	0	0	X	X	X	X
1	0	0	1	X	X	X	X
1	0	1	0	X	X	X	X
1	0	1	1	X	X	X	X
1	1	0	0	X	X	X	X
1	1	0	1	X	X	X	X
1	1	1	0	X	X	X	X
1	1	1	1	X	X	X	X

K-map for T_A is given in Figure 9.21.

AB/CD	00	01	11	10
00	x	1	x	
01		x	x	x
11	x	x	x	x
10		x	x	x

Figure 9.21 K-map for T_A.

$$T_A = D$$

K-map for T_B is given in Figure 9.22.

AB/CD	00	01	11	10
00	x		x	
01		x	x	x
11	x	x	x	x
10	1	x	x	x

Figure 9.22 K-map for T_B.

$$T_B = A$$

K-map for T_C is given in Figure 9.23.

AB/CD	00	01	11	10
00	x		x	
01	1	x	x	x
11	x	x	x	x
10		x	x	x

Figure 9.23 K-map for T_C.

$$T_C = B$$

K-map for T_D is given in Figure 9.24.

AB/CD	00	01	11	10
00	x		x	1
01		x	x	x
11	x	x	x	x
10		x	x	x

Figure 9.24 K-map for T_D.

$$T_D = C$$

The four stage ring counter logical circuit is given in Figure 9.25.

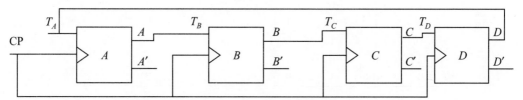

Figure 9.25 Four stage ring counter circuit diagram.

9.8 JOHNSON COUNTER

The Johnson counter is additionally called the twisted-ring counter. This counter is strictly identical because the ring counter except the inverted output of the last flip flop is connected to the input of the primary flip flop.

Let us the number started from 000, 100, 110, 111, 011 and 001, and also the sequence is continual goodbye as there is input pulse.

Like tally or rotating knowledge around a continual loop, ring counters may also be wont to notice or acknowledge numerous patterns or range values inside a group of information. By connecting easy logic gates like the AND or the OR gates to the outputs of the flip flops the circuit will be created to notice a group range or price. Standard 2, 3 or 4-stage Johnson ring counters may also be wont to divide the frequency of the clock signal by variable their feedback connections and divide-by-3 or divide-by-5 outputs are accessible. Let us consider four-bit Johnson control T Type flip flop. Four bits mean four flip flops are required to design the Johnson counter. The state diagram of Johnson counter is given in Figure 9.26.

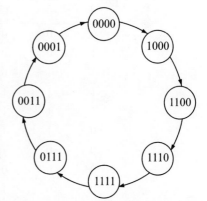

Figure 9.26 Johnson counter state diagram.

224 Digital Logic Design

The state table of Johnson counter is given in Table 9.11.

Table 9.11 Johnson counter state table

Present State				Next State			
A	B	C	D	A	B	C	D
0	0	0	0	1	0	0	0
1	0	0	0	1	1	0	0
1	1	0	0	1	1	1	0
1	1	1	0	1	1	1	1
1	1	1	1	0	1	1	1
0	1	1	1	0	0	1	1
0	0	1	1	0	0	0	1
0	0	0	1	0	0	0	0
0	0	1	0	?	?	?	?
0	1	0	0	?	?	?	?
0	1	0	1	?	?	?	?
0	1	1	0	?	?	?	?
1	0	0	1	?	?	?	?
1	0	1	0	?	?	?	?

The excitation table is given in Table 9.12.

Table 9.12 Johnson counter excitation table

Present state				Flip Flop Inputs			
A	B	C	D	T_A	T_B	T_C	T_D
0	0	0	0	1	0	0	0
1	0	0	0	1	1	0	0
1	1	0	0	1	1	1	0
1	1	1	0	1	1	1	1
1	1	1	1	0	1	1	1
0	1	1	1	0	0	1	1
0	0	1	1	0	0	0	1
0	0	0	1	0	0	0	0
0	0	1	0	X	X	X	X
0	1	0	0	X	X	X	X
0	1	0	1	X	X	X	X
0	1	1	0	X	X	X	X
1	0	0	1	X	X	X	X
1	0	1	0	X	X	X	X

K-map for T_A is given in Figure 9.27.

AB/CD	00	01	11	10
00	1			x
01	x	x		x
11	1	x		1
10	1	x	x	x

Figure 9.27 K-map for T_A.

$$T_A = D'$$

K-map for T_B is given in Figure 9.28.

AB/CD	00	01	11	10
00				x
01	x	x		x
11	1	x	1	1
10	1	x	x	x

Figure 9.28 K-map for T_B.

$$T_B = A$$

K-map for T_C is given in Figure 9.29.

AB/CD	00	01	11	10
00				x
01	x	x	1	x
11	1	x	1	1
10		x	x	x

Figure 9.29 K-map for T_C.

$$T_C = B$$

K-map for T_D is given in Figure 9.30.

AB/CD	00	01	11	10
00			1	x
01	x	x	1	x
11		x	1	1
10		x	x	x

Figure 9.30 K-map for T_D.

$$T_D = C$$

The Johnson counter logical circuit is given in Figure 9.31.

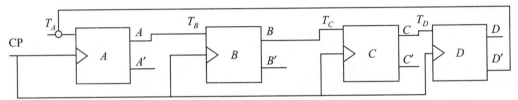

Figure 9.31 Johnson counter logical circuit diagram.

9.9 RIPPLE COUNTER

A ripple counter is an associate degree asynchronous counter wherever solely the primary flip flop is clocked by associate degree of external clock. All future flip flops are clocked by the output of the preceding flip flop. Asynchronous counters are known as ripple-counters owing to the manner the clock pulse ripples it through the flip flops. In another word we can say the initial clock pulse is given to first flip flop and the clock pulse is given to second flip flop from first flip flop and so on. Let us design a three-bit binary ripple counter or asynchronous counter using T type flip flop. Three-bit counter means three flip flops are required to design the ripple or asynchronous counter. The excitation table for three binary counters is given in Table 9.13.

Table 9.13 Three bits binary ripple counter excitation table

A	B	C	T_A	T_B	T_C
0	0	0	0	0	1
0	0	1	0	1	1
0	1	0	0	0	1
0	1	1	1	1	1
1	0	0	0	0	1
1	0	1	0	1	1
1	1	0	0	0	1
1	1	1	1	1	1

K-map for T_A is given in Figure 9.32.

AB/C	0	1
00		
01		1
11		1
10		

Figure 9.32 K-map for T_A.

$$T_A = BC$$

K-map for T_B is given in Figure 9.33.

AB/C	0	1
00		1
01		1
11		1
10		1

Figure 9.33 K-map for T_B.

$$T_B = C$$

K-map for T_C is given in Figure 9.34.

AB/C	0	1
00	1	1
01	1	1
11	1	1
10	1	1

Figure 9.34 K-map for T_C.

$$T_C = 1$$

The three bits binary ripple counter logical circuit is given in Figure 9.35.

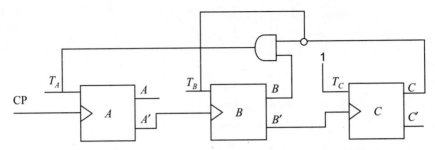

Figure 9.35 Three bits binary ripple counter logical circuit diagram.

Figure 9.35 shows the three bits Down ripple counter. Figure 9.36 shows the three bits Up ripple counter.

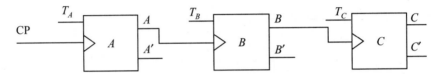

Figure 9.36 Three bits down ripple counter logical circuit.

EXERCISES

Short Answer Questions

1. What is counter?
2. Why is counter used?
3. What is BCD counter?
4. What is synchronous counter?
5. What is asynchronous counter?
6. What is ripple counter?
7. What is Johnson counter
8. What is difference between asynchronous and synchronous counters?
9. What is decade counter?
10. What is ring counter?
11. What is the purpose of counter?
12. How many inputs are required to design a counter to count a sequence 4,5,8,2,1,5.
13. How many flip flops are required to design ring counter?
14. How many flip flops are required to design BCD counter?

Long Answer Questions

1. Design the counter to count the following sequence using JK flip flop.
   (a) 0,1,2,3,5,6
   (b) 1,3,5,6,8,9
   (c) 2,4,8,10,12,14
   (d) 1,1,2,2,0,3,4,5,7,9
   (e) 3,5,7,9,11,13,15
2. Design a four bits up counter using RS flip flops.
3. Design a three bits binary ripple counter using D type of flip flops.
4. Design three bits synchronous ring counter using T type flip flops.
5. Design the synchronous counter from the following given state table using T type of flip flop.

Present State			Next State		
X	Y	Z	X	Y	Z
0	0	0	0	0	0
0	0	1	1	0	0
0	1	0	1	1	0
0	1	1	1	1	1
1	0	0	1	1	1
1	0	1	0	1	1
1	1	0	0	0	1
1	1	1	0	0	0

6. Design an asynchronous counter for counting the sequence 1, 1, 2, 2, 3, 4, 5, 6, 7 using D type flip flop.
7. Design a synchronous counter from the following state diagram using JK flip flop.

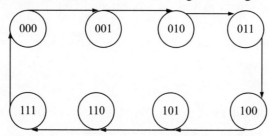

8. Design the parallel counter from the following diagram using RS.

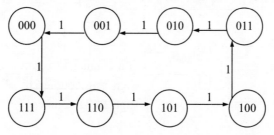

9. What is ripple counter? Design a four-bit binary ripple counter using T type flip flop.
10. Draw the state table from the following excitation table of a counter.

A	B	C	T_A	T_B	T_C
0	0	0	0	0	1
0	0	1	0	1	1
0	1	0	0	0	1
0	1	1	1	1	1
1	0	0	0	0	1
1	0	1	0	1	1
1	1	0	0	0	1
1	1	1	1	1	1

11. Draw the state diagram from state table obtained in Exercise 8.12.
12. Design the counter from the excitation table given in Exercise 8.12.
13. Design the counter to count the following sequence using T type flip flop.
    (a) 0,2,4,6,8,10,12,14
    (b) 1,3,5,7,9,11,13
    (c) 1,2,3,4,5,4,3,2,1
    (d) 0,1,2,5,6,7,11,12,13,15,16,17
    (e) 10,9,8,7,6,5,4,3,2,1
14. Differentiate between synchronous counter and asynchronous counter.
15. What is Johnson counter? Design three bits synchronous Jonson counter.

Multiple Choice Questions

1. Counter is a
   (a) Combinational circuit
   (b) Sequential circuit
   (c) Combinational circuit as well as sequential circuit
   (d) None
2. How many flip flops are required to design a counter to count the following sequence 0, 1, 2, 6, 3?
   (a) 3
   (b) 4
   (c) 5
   (d) 2
3. The three-bit counter is given below. What is the name of this counter?

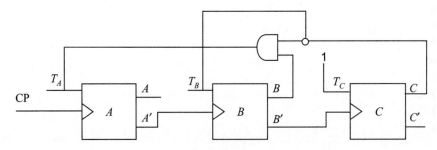

(a) Up counter (b) Down counter
(c) Ripple counter (d) Binary counter
4. Which of the following counter is additionally called the twisted-ring counter?
   (a) Johnson counter (b) Down counter
   (c) Ripple counter (d) Binary counter
5. Which of the following is Johnson counter state diagram?

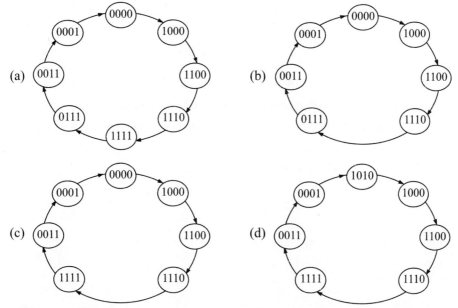

6. How many flip flops are required to design synchronous BCD counter?
   (a) 3 (b) 4
   (c) 5 (d) 6
7. How many JK flip flops are required to design mod-8 counter?
   (a) 3 (b) 4
   (c) 5 (d) 6
8. For every _____ the final output of a mod-6 counter occurs one time.
   (a) 7 clock pulse (b) 6 clock pulse
   (c) 5 clock pulse (d) 4 clock pulse
9. Six (6) bits synchronous counter has maximum number of modulus is:
   (a) 36 (b) 32
   (c) 64 (d) 16
10. The Up/Down counter works as
    (a) UP count active-High and the down count active-Low
    (b) UP count active-High and the down count active-High
    (c) The down count active-High and Up count active-Low
    (d) The down count active-High and Up count active-High

11. Decade counter is used to count from
    (a) 0000 to 1111
    (b) 1111 to 0000
    (c) 0011 to 1111
    (d) 0001 to 1111

Answers

| 1. (b) | 2. (b) | 3. (c) | 4. (a) | 5. (a) | 6. (b) |
| 7. (a) | 8. (b) | 9. (b) | 10. (a) | | |

BIBLIOGRAPHY

http://www.daenotes.com/electronics/digital-electronics/counters-types-of-counters.

http://www.doc.ic.ac.uk/~nd/surprise_96/journal/vol4/cwl3/report.html#bcd.

http://people.wallawalla.edu/~curt.nelson/engr354/lecture/brown/chapter7_reg_counters.pdf.

http://www.electronics-tutorials.ws.

http://electronics-course.com/ripple-counter.

http://www.indiabix.com/.

CHAPTER 10

Register Design

10.1 INTRODUCTION

Register is a group of flip flops. Four-bit register means four flip flops are grouped together to store four-bits information. The register is example of sequential circuit. The registers are used frequently to make huge sequential circuits for different purposes. They can hold the large amount of data rather than individual flip flops. The register is an important component to design the modern processors. There are various kinds of register available in the real world. There are also many applications of register. In other words you can say register is extension of flip flop to store many bits. The registers are generally used as a temporary storage in processors. The registers are faster than main memory and also convenient. The register is very helpful to speed up the complex calculation. Consider an example of two-bit register. The two-bit register required two flip flops. This two-bit register stores two-bit data information. The diagram of two-bit register is given in Figure 10.1.

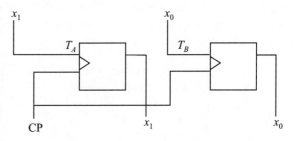

Figure 10.1 Two-bit register diagram.

The flip flop is treated as one cell. The block diagram is given in Figure 10.2.

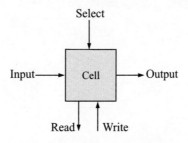

Figure 10.2 Cell block diagram.

Select : It is used to select the particular cell in which data is to write or read in the register.
Input : It is used to input the data/information for storage in particular cell in the register.
Output : It is used to get data from a particular cell from the register.
Read : This line is used for sending the read signal to read the data from a particular cell in the register.
Write : This line is used to send a signal for writing data in a particular cell in the register.

10.2 TYPES OF REGISTER

There are various types of register available in the real world. The various kinds of register are differentiated according to their use and capabilities. The various types of register are Memory Buffer Register (MBR), Memory Address Register (MAR), Data Register (DR), Instruction Register (IR), Stack Pointer (SP), General Purpose Register (GPR), etc.

10.3 ADDING A PARALLEL LOAD OPERATION

Let us consider the four-bit register in which four input X_3, X_2, X_1 and X_0 are copired into the output Y_3, Y_2, Y_1 and Y_0 on each clock pulse. The problem is here that how can you store the current value on more than one clock pulses. This problem can be solved by adding a load input signal L to the register. If $L = 0$, then the register C holds its current value and if $L = 1$ then the register stores a new value, which is taken from the inputs X_3, X_2, X_1 and X_0. The truth table for this problem is given in Table 10.1.

Table 10.1 Four bits register truth table

L	Y(t + 1)
0	Y(t)
1	X_3, X_2, X_1 and X_0

The block diagram is given in Figure 10.3.

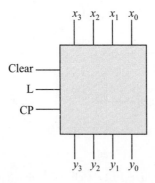

Figure 10.3 Four bits register block diagram.

10.4 REGISTER WITH PARALLEL LOAD

The register with parallel load, in which if load $L = 0$ then the flip flop inputs are X_3, X_2, X_1 and X_0. Therefore, every flip flop just holds its current value. When the load $L = 1$ then the flip flop inputs are Y_3, Y_2, Y_1 and Y_0. Then, this new value is added into the register. The diagram of register with parallel load is given in Figure 10.4.

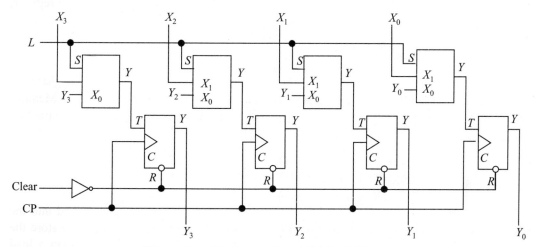

Figure 10.4 Register with parallel load diagram.

10.5 SHIFT REGISTER

The shift registers are those registers which have ability to shift the content of one register to another register. In other words you can say a shift register "shifts" its output once per clock cycle. The groups of flip flops are connected in a chain in order that the output from one flip flop becomes the input of succeeding flip flops. Most of the registers possess no characteristic internal sequence of states. All the flip flops are driven by a standard clock, and every one area unit set or reset at the same time. The various types of shift register are

- Serial In Serial Out (SISO)
- Serial In Parallel Out (SIPO)
- Parallel In Serial Out (PISO)
- Parallel In Parallel Out (PIPO)
- Bidirectional Shift Registers (BSR)

The block diagram of shift register is given in Figure 10.5.

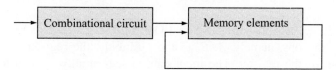

Figure 10.5 Shift register block diagram.

236 Digital Logic Design

Let us consider four bits shift register which is constructed with four T type flip flops A, B, C and D. The four-bits shift register using T type flip flops is given in Figure 10.6.

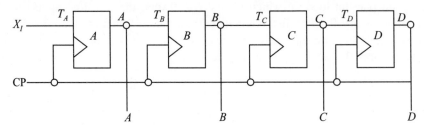

Figure 10.6 Four bits shift register.

X_I is a new input bit supply to T_A and it will be shifted to next flip flop after a clock pulse. Therefore, the next states of flip flops A, B, C and D are given below.

$$A(t + 1) = X_I$$
$$B(t + 1) = A(t + 1)$$
$$C(t + 1) = B(t + 1)$$
$$D(t + 1) = C(t + 1)$$

EXAMPLE 10.1 The contents of register X is 1101, shift the contents of register X to register Y.

Solution: $X = 1101$
$Y = 0000$

CP	X				Y			
0	1	1	0	1	0	0	0	0
1	1	1	1	0	1	0	0	0
2	0	1	1	1	0	1	0	0
3	1	0	1	1	1	0	1	1
4	1	1	0	1	1	1	0	1

In the first clock pulse (CP) the last bit of X is shifted to first bit position register Y and the same bit inserted at the beginning of X register. In the 2nd clock pulse the second last bit of register X shifted to first bit position of register Y and the 1st bit of register Y shifted to 2nd bit position in Y register, the second 1st bit also inserted to the beginning of register X and other bit of register shifted accordingly and in this way all bits shifted from register X to register Y. The four-bit register required four clock pulses to shift its contents.

10.5.1 Serial In Serial Out (SISO) Shift Register

Let us consider a basic four-bit Serial In Serial Out Shift Register. This register is constructed with four T type flip flops, namely, A, B, C and D. Firstly, the register is cleared by forcing zero to four output of the register. The input data is sequentially applied to the input of the first flip flop A from the left side of the flip flop. On every clock pulse each bit is shifted from left flip flop to the right flip flop. The content of the register is shifted out at the last clock pulse and the contents of register become zero. The diagram of Serial In Serial Out (SISO) Shift Register is given in Figure 10.7.

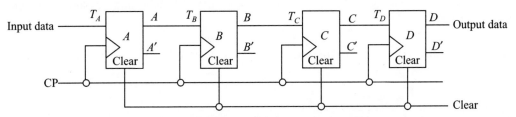

Figure 10.7 Serial In Serial Out (SISO) shift register diagram.

T_A = data, $T_B = A$, $T_C = B$, and $T_D = C$

EXAMPLE 10.2 The contents of four-bit register X is 1001, shift Serial In Serial Out (SISO) the contents of register X.

Solution: $X = 1001$

Clock Pulse	Register X				Output			
0	1	0	0	1				
1	0	1	0	0	0			
2	0	0	1	0	0	0		
3	0	1	0	0	0	0	1	
4	0	0	0	0	1	0	0	1

10.5.2 Serial In Parallel Out (SIPO) Shift Register

Serial In Parallel Out Shift Register accepts input data in serial and produces output in parallel. The block diagram of this type of shift register is given in Figure 10.8. The given shift register is constructed with four T type flip flops A, B, C and D. This register stores four-bit data. This shift register is also used to convert the data given in serial data form to parallel data form. This is also used where data coming from a single wire and you want to route on parallel line. The Serial In Parallel Out (SIPO) shift register is given in Figure 10.8.

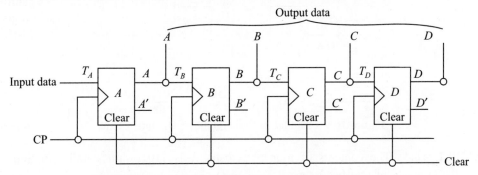

Figure 10.8 Serial In Parallel Out (SIPO) shift register.

Consider X is a four-bit data stream, $X = X_0X_1X_2X_3$ as input data used in serial in parallel out shift register. Let Y is output data stream of serial in parallel out shift register, the bits of Y are Y_0, Y_1, Y_2 and Y_3, so $Y = Y_0Y_1Y_2Y_3$. The Y_0 bit of Y is replaced with X_0 on first clock pulse. The bit Y_1 of Y is replaced with X_1 of X on second clock pulse and similarly it continues till the last bit of X.

EXAMPLE 10.3 The contents of four-bit register X is 1001, shift Serial In Parallel Out (SIPO) the contents of register X.

Solution: $X = 1011$

Clock Pulse	Register X				Output			
0	1	0	1	1				
1	0	0	1	1	1			
2	0	0	1	1	1	0		
3	0	0	0	1	1	0	1	
4	0	0	0	0	1	0	1	1

10.5.3 Parallel In Serial Out (PISO) Shift Register

Data is given in parallel to Parallel In Serial Out Shift Register and output is produced in serial. This register is reverse of Serial In Parallel Out Shift Register. In parallel in serial out shift register you can load data in parallel into all stages before any shifting starts. Parallel in serial out is used to convert data from a parallel data format to a serial data format. Parallel data format means that the presence of data bits on individual wires simultaneously. The serial data format means that the bits data are sequentially presented on a single wire in time. This register is designed with four flip flops, namely, A, B, C and D. The T type flip flops are used to design this shift register. The four clock pulses are required to shift the input data to produce output in serial data format. The diagram of Parallel In Serial Out is given in Figure 10.9.

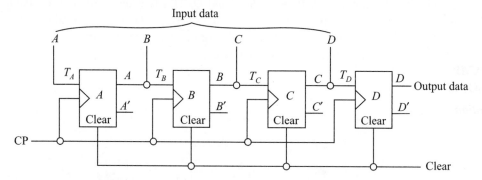

Figure 10.9 Parallel In Serial Out (PISO) shift register.

EXAMPLE 10.4 The contents of four-bit register X is 1101, shift Parallel In Serial Out (PISO) the contents of register X.

Solution:

Clock Pulse	Register X				Output			
0	1	1	0	1				
1	1	1	1	0	1			
2	1	1	1	1	0	1		
3	1	1	1	1	1	0	1	
4	0	0	0	0	1	1	0	1

10.5.4 Parallel In Parallel Out (PIPO) Shift Register

The Parallel In Parallel Out Shift Register is used to take data in parallel and shifts it to output in parallel. The four data bits, A, B, C and D, are given to four-bit parallel in parallel out shift register at T_A, T_B, T_C and T_D. The output is also produced in parallel as A, B, C and D. The data bit is shifted in one bit position on every clock pulse. The diagram of Parallel In Parallel Out (PIPO) shift register is given in Figure 10.10.

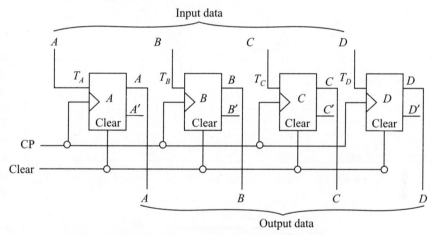

Figure 10.10 Parallel In Parallel Out (PIPO) shift register.

EXAMPLE 10.5 The contents of four-bit register X is 1011, shift Parallel In Parallel Out (PISO) the contents of register X.

Solution:

Clock Pulse	Register X				Output			
0	1	0	1	1				
1	0	1	0	1	1			
2	0	0	1	0	1	1		
3	0	0	0	1	0	1	1	
4	0	0	0	0	1	0	1	1

10.5.5 Bidirectional Shift Register (BSR)

The data can be shifted either left or right in Bidirectional Shift Register. The logical gates are used to implement Bidirectional Shift Register that enables the transfer of a bit of data from one stage to the next stage either right or left, depending on the clock pulse or control signal. The Bidirectional Shift Register (BSR) is given in Figure 10.11.

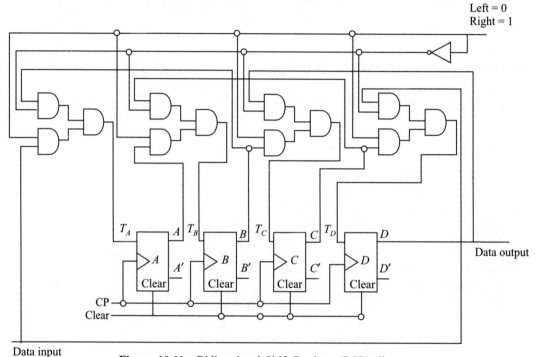

Figure 10.11 Bidirectional Shift Register (BSR) diagram.

EXERCISES

Short Answer Questions

1. What is register?
2. Explain purposes of register.
3. What are components of register?
4. What is PISO shift register?
5. What is SISO shift register?
6. How SISO shift register differ from PISO shift register?
7. What is Bidirectional Shift Register (BSR)?
8. What is shift register?
9. What are various types of register?
10. What do you mean by parallel load operation?

Long Answer Questions

1. Design six-bit serial in and serial out shift register.
2. What are various types of shift register? Explain any one with a suitable example.
3. What is bidirectional shift register? Explain with a suitable example.
4. The bit 1011 is serially entered from the right most bit first into a 6-bit parallel out shift register. The register is clear in starting. What are outputs after four clock pulses?
5. The bits 111011 are shifted in serial (right most bit first) into a 6-bit serial input and parallel output shift register. The initial state is 01110.
6. What are the contents of register after four clock pulses?
7. Let a 4-bit serial in and parallel out shift register is initially zero. You want to store the data bits 1100. After three clock pulses, write the four-bit pattern (right most bit first).
8. How many clock pulses will be used to fully load serially an 8-bit shift register?
9. The contents of register R are 11010. Shift the contents of register R to S.

Multiple Choice Questions

1. Four-bit register means:
   (a) 4 bits information is stored
   (b) 3 bits information is stored
   (c) 2 bits information is stored
   (d) 1 bit information is stored
2. The data can be shifted either left or right in _____ shift register.
   (a) Parallel In Parallel Out
   (b) Bidirectional
   (c) Parallel In Serial Out
   (d) Serial In Parallel Out
3. The circuit given below is a

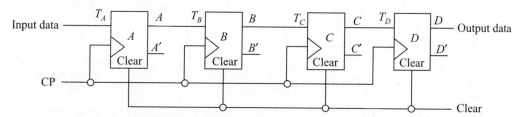

   (a) Parallel In Parallel Out Shift Register
   (b) Bidirectional Shift Register
   (c) Serial In Serial Out Shift Register
   (d) Serial In Parallel Out Shift Register
4. The contents of four-bit register X is 1011, shift Parallel In Parallel Out (PISO), the contents of register X is

   (a) | 1 | 0 | 1 | 1 |
   (b) | 1 | 0 | 0 | 1 |
   (c) | 1 | 0 | 1 | 0 |
   (d) | 1 | 0 | 0 | 0 |

242 Digital Logic Design

5. PIPO stands for
   (a) Parallel In Parallel Out
   (b) Parallel In Parallel Omit
   (c) Path In Parallel Out
   (d) Pipeline In Parallel Out
6. Which of the following is not the characteristic of shift register?
   (a) Parallel In and Parallel Out
   (b) Parallel In and Serial In
   (c) Parallel In and Serial Out
   (d) Serial In and Parallel Out
7. Let us consider a 4-bit serial in and serial out shift register in which initial value is zero. You want to store data 1000. What is the bit patter after second clock pulses?
   (a) 1100
   (b) 0011
   (c) 0000
   (d) 1001
8. The four-bit data is 1001 in a parallel in and parallel out shift register. After four clock pulses, the data outputs are:
   (a) 0101
   (b) 0000
   (c) 1100
   (d) 1101
9. How many clock pulses are required to complete load serially a 6 bits shift register?
   (a) 3
   (b) 4
   (c) 5
   (d) 6
10. What is the storage capacity on each stage in a shift register presentation?
    (a) 4 bits
    (b) 3 bits
    (c) 1 bit
    (d) 6 bits
11. The following circuit diagram represents the _____ type of shift register.

Left = 0
Right = 1

Data input

(a) Bidirectional
(b) Serial In Serial Out Shift Register
(c) Serial In Parallel Out Shift Register
(d) Parallel In Serial Out Shift Register

Answers

1. (a) 2. (b) 3. (c) 4. (a) 5. (b) 6. (c)
7. (b) 8. (d) 9. (a) 10. (a)

BIBLIOGRAPHY

http://www.ee.usyd.edu.au/tutorials/digital_tutorial/part2/register01.html.
http://www.web-books.com/eLibrary/Engineering/Circuits/Digital/DIGI_12P3.htm.
http://www.allaboutcircuits.com/vol_4/chpt_12/5.html.

CHAPTER 11

Threshold Circuit and Digital Computer Design

11.1 THRESHOLD LOGIC CIRCUIT

The logic circuit designed using threshold gate is termed as threshold logic circuit. The threshold gate is also known as T-gate. T-gate is more powerful than the conventional gates. T-gate is an element of threshold logic circuit.

11.1.1 Threshold Gate (T-gate)

The threshold gates are just like conventional gates which are used in conventional digital logic circuit. The T-gates are used to design the threshold circuit. The threshold gate accepts n binary inputs and produces one single binary output. There are two additional parameters added in this gate, namely, threshold T and weight W. Let us consider T gate with n inputs like $a_1, a_2, a_3, \ldots, a_n$ and weights $w_1, w_2, w_3, \ldots, w_n$. The weights w_1, w_2 and so on are associated with inputs $a_1, a_2,$ etc. The values of weight W and threshold T may be real, finite, negative or positive numbers. The thresholds gate is shown in Figure 11.1. For example, let us consider $a_1, a_2, a_3, \ldots, a_n$ are inputs and $w_1, w_2, w_3, \ldots, w_n$ are weights, so the threshold gate is defined as

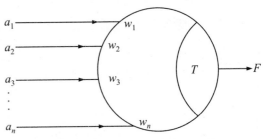

Figure 11.1 Threshold gate T.

$F(a_1, a_2, a_3, \ldots, a_n) = 1$ if $\sum_{i=1}^{n} a_i w_i \geq T$, where a_i are inputs, w_i are weights and T is the threshold gate.

And in other case
$$F(a_1, a_2, a_3, \ldots, a_n) = 0$$

EXAMPLE 11.1 Find the Boolean function from the following threshold gate.

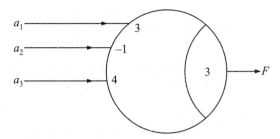

Solution: In the figure there are three inputs, namely, a_1, a_2 and a_3 and weights are 3, −1 and 4, respectively. Let the weighted sum is w. The value of w is calculated with the help of following equation:

$$w = w_1 a_1 + w_2 a_2 + w_3 a_3$$

Put the value of a_1, a_2, a_3, w_1, w_2 and w_3 in the above equation to get w.

$$w = 3a_1 + (-1)a_2 + 4a_3$$
$$w = 3a_1 - a_2 + 4a_3$$

$T = 3$ is given in the figure. If $(3a_1 - a_2 + 4a_3) \geq 3$, the output will be 1, otherwise 0 (zero). The input/output (I/O) relation is given in the following table.

Inputs			Weighted Sum	Output
a_1	a_2	a_3	$w = 3a_1 - a_2 + 4a_3$	F
0	0	0	0	0
0	0	1	4	1
0	1	0	−1	0
0	1	1	3	1
1	0	0	3	1
1	0	1	7	1
1	1	0	2	0
1	1	1	6	1

From the above table the Boolean function for output F is

$$F(a_1, a_2, a_3) = \sum (1, 3, 4, 5, 7)$$

EXAMPLE 11.2 Draw the threshold gate from the Boolean function $F(a, b, c) = \sum (2, 3, 5, 6, 7)$. The input variables of the given function are associated with the weights 2, −1 and 5, respectively.

Solution: Here a, b and c are three input variables and given weighs associated with inputs are 2, −1 and 5, respectively. The weighted sum $w = 2a - 1b + 5c$. The I/O relation is given in the following table.

Inputs			Weighted Sum	Output
a	b	c	$w = 2a - 1b + 5c$	F
0	0	0	0	0
0	0	1	5	1
0	1	0	−1	0
0	1	1	4	1
1	0	0	2	1
1	0	1	7	1
1	1	0	1	1
1	1	1	6	1

In the above table the minimum weighted sum is 1, therefore the value of $T = 1$. The threshold gate is given below.

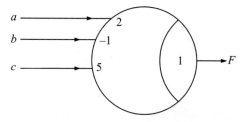

EXAMPLE 11.3 Find Boolean function from the threshold circuit in which there are four inputs a, b, c and d. The weights 3, 4, −2 and 6 are given which are associated with inputs, respectively. The value of T-gate is 4.

Solution: The inputs are a, b, c and d. The weight associated with inputs are 3, 4, −2 and 6, so the weighted sum $w = 3a + 4b - 2c + 6d$. The output is 1 (one) when $w \geq T$, otherwise 0 (zero). The I/O relation table is given below.

Inputs				Weighted Sum	Output
a	b	c	d	$w = 3a + 4b - 2c + 6d$	F
0	0	0	0	0	0
0	0	0	1	6	1
0	0	1	0	−2	0
0	0	1	1	4	1
0	1	0	0	4	1
0	1	0	1	10	1
0	1	1	0	2	0
0	1	1	1	8	1
1	0	0	0	3	0
1	0	0	1	9	1
1	0	1	0	1	0
1	0	1	1	7	1
1	1	0	0	7	1
1	1	0	1	13	1
1	1	1	0	5	1
1	1	1	1	11	1

Function is $F(a, b, c, d) = \sum(1, 3, 4, 5, 7, 9, 11, 12, 13, 14, 15)$

11.1.2 Implementation of Conventional Gates with T-gate

The conventional gate AND, OR and NOT can be implemented with T-gate or threshold gate. The universal gates NAND and NOR can also be implemented with T-gate.

AND gate implementation with T-gate

The following threshold gate or T-gate is equivalent to AND gate. The T-gate equivalent to AND gate is given in Figure 11.2.

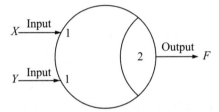

Figure 11.2 T-gate equivalent to AND gate.

The I/O relation table for the above T-gate is given in Table 11.1.

Table 11.1 I/O relation table for T-gate in Figure 11.2

Inputs		Weighted Sum	Output
X	Y	w = X + Y	F
0	0	0	0
0	1	1	0
1	0	1	0
1	1	2	1

The output function F is $F(X, Y) = X.Y$ which is equation of AND gate, so you can use the above T-gate to replace AND. The above T-gate will work as AND gate.

OR gate implementation with T-gate

The given below T-gate is equivalent to conventional OR gate. The output of the given T-gate is equal to the output of OR gate. The T-gate equivalent to OR gate is given in Figure 11.3.

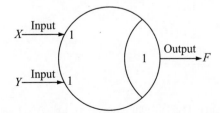

Figure 11.3 T-gate equivalent to OR gate.

The I/O relation table for the above T-gate is given in Table 11.2.

Table 11.2 I/O relation table for T-gate in Figure 11.3

Inputs		Weighted Sum	Output
X	Y	$w = X + Y$	F
0	0	0	0
0	1	1	1
1	0	1	1
1	1	2	1

The output function F is
$F(X, Y) = X'Y + XY' + XY = X'Y + X(Y' + Y) = X'Y + X = X + Y$, so $F(X, Y) = X + Y$, which is the equation of OR gate, so you can use the above T-gate to replace OR. The above T-gate will work as OR gate.

NOT gate implementation with T-gate

The given below T-gate is equivalent to conventional NOT gate. The output of the given T-gate is equal to the output of NOT gate. The T-gate equivalent to NOT gate is given in Figure 11.4.

The I/O relation table for the above T-gate is given in Table 11.3.

Table 11.3 I/O relation table for T-gate in Figure 11.4

Input	Weighted Sum	Output
X	$w = -X$	F
0	0	0
1	−1	1

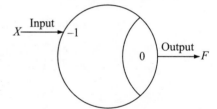

Figure 11.4 T-gate equivalent to NOT gate.

The output function F is $F(X) = -X$, which is the equation of NOT gate, so you can use the above T-gate to replace NOT. The above T-gate will work as NOT gate.

NAND gate implementation with T-gate

The given below T-gate is equivalent to universal gate NAND gate. The output of the given T-gate is equal to the output of NAND gate. The T-gate equivalent to NAND gate is given in Figure 11.5.

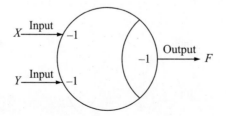

Figure 11.5 T-gate equivalent to NAND gate.

The I/O relation table for the above T-gate is given in Table 11.4.

Table 11.4 I/O relation table for T-gate in Figure 11.5

Inputs		Weighted Sum $w = -X + (-)Y$ $w = -X - Y$	Output F
X	Y		
0	0	0	1
0	1	-1	1
1	0	-1	1
1	1	-2	0

The output function F is
$F(X) = X'Y' + X'Y + XY' = X'(Y' + Y) + XY' = X' + XY' = X' + Y' = (XY)'$, which is the equation of NAND gate, so you can use the above T-gate to replace NAND. The above T-gate will work as NAND gate.

NOR gate implementation with T-gate

The T-gate given below is equivalent to universal gate NOR gate. The output of the given T-gate is equal to the output of NOR gate. The T-gate equivalent to NOR gate is given in Figure 11.6.

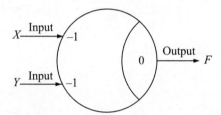

Figure 11.6 T-gate equivalent to NOR gate.

The I/O relation table for the above T-gate is given in Table 11.5.

Table 11.5 I/O relation table for T-gate in Figure 11.6

Inputs		Weighted Sum $w = -X - Y$	Output F
X	Y		
0	0	0	1
0	1	-1	0
1	0	-1	0
1	1	-2	0

The output function F is $F(X) = X'Y' = (X + Y)'$, which is the equation of NOR gate, so you can use the above T-gate to replace NOR. The above T-gate will work as NOR gate.

EXAMPLE 11.4 Implement the following Boolean function with T-gate.

$$F(X,Y) = \sum(1, 2)$$

Solution: The conventional circuit diagram of the Boolean function $F(X, Y) = \sum(1, 2)$ is given below. The Boolean function $F(X, Y) = \sum(1, 2)$ can be written as $F(X, Y) = X'Y + XY'$

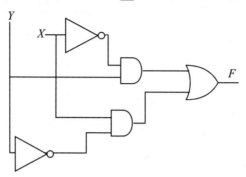

Replace all conventional gates with equivalent T-gate in the above circuit diagram to get threshold logic circuit for the above conventional logic circuit diagram.

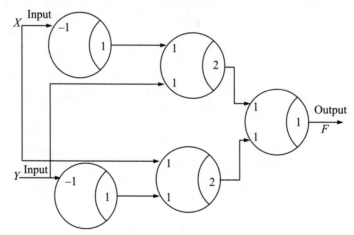

EXAMPLE 11.5 Implement the following conventional logic circuit with T-gate.

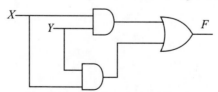

Solution: Replace AND and OR gate with T-gate to get the desired result.

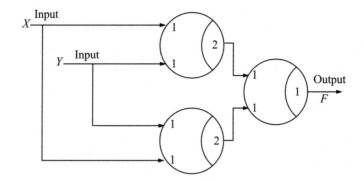

EXAMPLE 11.6 Find the Boolean function from the following threshold logic circuit.

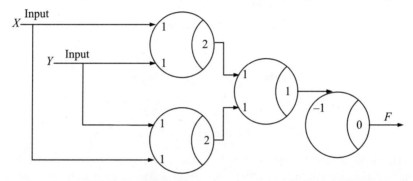

Solution: Let $F1$ is the output of first gate and $F2$ is the output of second gate. These two $F1$ and $F2$ become the inputs of third gate. The $F3$ is the output of third gate and becomes the input for fourth gate and F is the final output.

Inputs		Weighted Sum		Outputs		Weighted Sum for F1 and F2	Output	Weighted Sum for F3	Final Output
X	Y	$T=2$, $w=X+Y$	$T=2$, $w=X+Y$	F1	F2	$w = F1 + F2$	F3	$w = -F3$	F
0	0	0	0	0	0	0	0	0	1
0	1	1	1	0	0	0	0	0	1
1	0	1	1	0	0	0	0	0	1
1	1	2	2	1	1	2	1	−1	0

The Boolean function from the above table is $F(X, Y) = \sum(0, 1, 2)$.

11.2 ARITHMETIC CIRCUIT DESIGN

CPU (Central Processing Unit) consists of ALU (Arithmetic Logic Unit), Registers and Control unit. ALU is responsible for performing arithmetic and logical operations. ALU consists of arithmetic circuit for performing arithmetic operations and logic circuit for performing logical operations.

11.2.1 Arithmetic Circuit

Arithmetic circuit is constructed for performing various arithmetic operations using the concept of design for combinational circuits. If conventional method of constructing arithmetic circuit is used then for designing even 4-bit arithmetic circuit for performing just two operation 29 minterms are to be generated and hence 9 variable K-maps is needed for simplification which becomes very tedious almost impossible. So to design such circuits divide and concur approach is employed. Larger circuit is broken into smaller circuits which can be constructed using conventional approach easily and then these smaller circuits are used to construct the larger one. For example to construct 4-bit arithmetic circuit 1-bit arithmetic circuit is constructed and then using these as building block 4-bit or more size arithmetic circuits are constructed. Constructions of arithmetic circuits through examples are given in Figure 11.7.

EXAMPLE 11.7 Construct a 4-bit arithmetic circuit to perform the function/operation on 4-bit operand A and B as per the following function table given below.

Control Signal $S1$	Function F
0	$A + B$
1	$A - B$

Solution: The 4-bit arithmetic circuit may be broken into four 1-bit arithmetic circuits. These four 1-bit arithmetic circuits may be connected in cascade to give 4-bit arithmetic circuit. In order to construct 1-bit arithmetic circuit standard circuit Full Adder (FA) may be used. FA is capable of adding three bits including two data bits and one carry bit. Using 2's complement concept subtraction may be converted into addition.

$$A - B = A + 2\text{'s complement of } B$$
$$= A + 1\text{'s complement of } B + 1$$
$$= A + B' + 1$$

So, modified function table is

Control Signal $S_i F_i$	Function
0	$A_i + B_i + 0$
1	$A_i + B_i' + 1$

Recall FA has got three inputs X_i, Y_i and C_i as well as two outputs S_i and $C_i + 1$ with $S_i = X_i + Y_i + C_i$. If F_i and S_i are compared term wise then

(i) $X_i = A_i$
(ii) $Y_i = f(B_i, S_1)$ = function of B_i and S_1 such that its value is B_i when S_1 is 0 and B_i' when S_i is 1 resulting into following truth table for the circuit computing Y_i as a function of S_1 and B_i

S_i	B_i	Y_i
0	0	0
0	1	1
1	0	1
1	1	0

(iii) $C_i = S_i$: The truth table of the circuit computing Y_i is nothing but Exclusive OR gate.

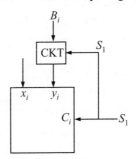

Hence 1-bit arithmetic circuit is

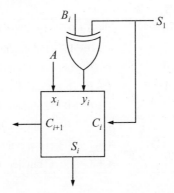

For constructing N bit arithmetic circuit N 1-bit arithmetic circuits are connected in cascade by connecting carry out of one arithmetic circuit to carry input of next 1-bit arithmetic circuit starting from LSB to MSB.

Thus 4 bits arithmetic circuit satisfying the given functionalities is

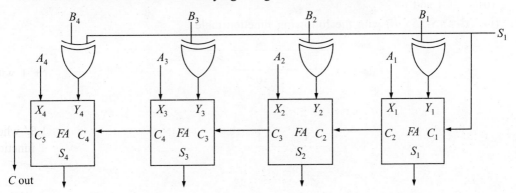

EXAMPLE 11.8 Design 4-bit arithmetic circuit for implementing the following function table.

Control signals			Function	
S_2	S_1	S_0	F	
0	0	0	A	(Transfer A)
0	0	1	$A + 1$	(Increment A)
0	1	0	$A + B$	(Add A and B)
0	1	1	$A + B + 1$	(Add A and B with carry)
1	0	0	$A - B - 1$	(Subtract B from A with borrow)
1	0	1	$A - B$	(Subtract B from A)
1	1	0	$A - 1$	(Decrement A)
1	1	1	A	(Transfer A)

Solution: The given function table may be rewritten as

Control signals			Function	
S_2	S_1	S_0	F	
0	0	0	$A + 0 + 0$	(Transfer A)
0	0	1	$A + 0 + 1$	(Increment A)
0	1	0	$A + B + 0$	(Add A and B)
0	1	1	$A + B + 1$	(Add A and B with carry)
1	0	0	$A + B' + 0$	(Subtract B from A with borrow)
1	0	1	$A + B' + 0$	(Subtract B from A)
1	1	0	$A + 1 + 0$	(Decrement A)
1	1	1	$A + 1 + 1$	(Transfer A)

Now let us construct 1-bit arithmetic circuit that implements the above functions. If we compare the functions expressed as sum of three terms with three inputs X, Y and C of Full Adder it can be easily derived that

(i) $C_0 = S_0$
(ii) $X = A$ and
(iii) $Y = f(S_2, S_1, B)$ with the following function table

S_2	S_1	Y
0	0	0
0	1	B
1	0	B'
1	1	1

Or the following K-map for Y

S_1 \ S_2	0	1
0		B
1	B'	1

From the above K-map or function table following truth table may be derived

Input			Output
S_2	S_1	B	Y
0	0	0	0
0	0	1	0
0	1	0	0
0	1	1	1
1	0	0	1
1	0	1	0
1	1	0	1
1	1	1	1

Using three variable K-map Y may be simplified as below.

S_1B \ S_2	0	1
00		1
01		
11	1	1
10		1

So, $Y = S_1B + S_2B'$

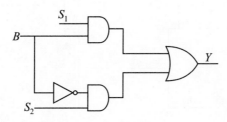

Using this expression a combinational circuit may be implemented to give Y_i from S_2, S_1 and B_i for all $i = 0$–3.

Using (i)–(iii) 1-bit arithmetic circuit should be as given below where S_2, S_1, S_0 are control signals and C_0 is input carry, C_1 is output carry and A_0, B_0 and F_0 are LSB of A, B and F, respectively.

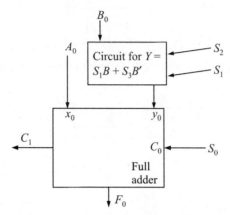

Four 1-bit arithmetic circuits may be connected in cascade with carry out of one fed into carry in of another one. Thus 4-bit arithmetic circuit designed is given below.

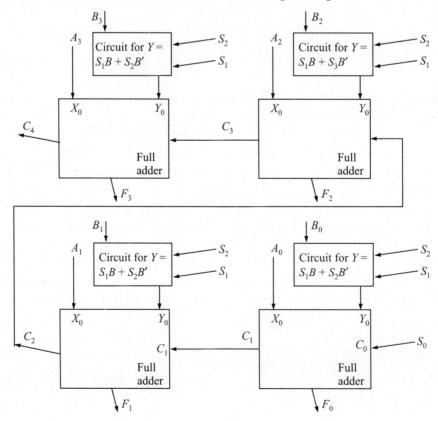

11.2.2 Logical Circuits

Logical circuits perform logical operation and can be designed using basic concepts of design of combinational circuits. Design of LC is illustrated through examples.

EXAMPLE 11.9 Design 4-bit logical circuit to perform as per function table given below.

Control Signals		Function
S_2	S_1	F
0	0	A OR B
0	1	A AND B
1	0	A EX-OR B
1	1	NOT A

Solution: In order to design 4-bit logical circuits 1-bit logical circuit is constructed and the same is replicated four time with four different input data set (A_i, B_i) for $i = 0 - 3$ giving four output bits F_i, $i = 0 - 3$.

The logical functions A_0 OR B_0, A_0 AND B_0, A_0 EX-OR B_0 and NOT A_0 may be generated with OR gate, NAND gate, EX-OR gate and NOT gate, respectively, and then any one of these depending on control signal may be available at output using 4:1 multiplexer.

Hence 1-bit logical circuit performing logical function on A_i and B_i is given below.

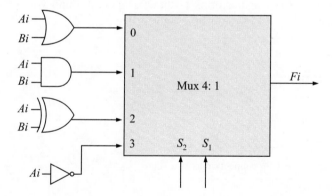

11.2.3 Arithmetic Logic Unit (ALU)

ALU performs both arithmetic and logic operations. Let us see the design of N-bit ALU through example. In order to design N-bit ALU 1-bit ALU is designed first and then the same N 1-bit ALUs are connected in cascade to give N-bit ALU.

EXAMPLE 11.10 Design 4-bit ALU with following functionalities

S_3	S_2	S_1	S_0	F	
0	0	0	0	A	(Transfer A)
0	0	0	1	$A + 1$	(Increment A)
0	0	1	0	$A + B$	(Add A and B)
0	0	1	1	$A + B + 1$	(Add A and B with carry)
0	1	0	0	$A - B - 1$	(Subtract B from A with borrow)
0	1	0	1	$A - B$	(Subtract B from A)
0	1	1	0	$A - 1$	(Decrement A)
0	1	1	1	A	(Transfer A)
1	0	0	X	A OR B	
1	0	1	X	A AND B	
1	1	0	X	A EX-OR B	
1	1	1	X	NOT A	

Solution: 4-bit ALU may be designed using 4 1-bit ALU. So let us design 1-bit ALU for the above functionalities.

1-bit ALU may be divided into 1-bit arithmetic circuit and 1-bit logic circuit whose design we have already seen in the previous sections. After carefully seeing the function table of ALU it is known that when S_3 is 0 it performs arithmetic operation and when S_3 is 1 it performs logic operation. Hence the ALU should be as given below.

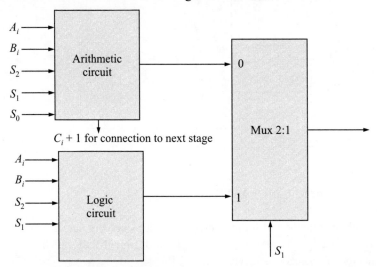

Arithmetic circuit and logic circuit are already given in the above figures.

Now connect 4 1-bit ALU in cascade to get 4-bit ALU. See the design of this 4-bit ALU using conventional method which was impossible as 12 input variables are, so 4K minterms and hence their listing and then simplification using 12 variable K-maps just impossible.

We may design the above ALU in another way as well. In the second way the arithmetic circuit is modified to perform all arithmetic and logic operations both and no separate logic circuit is constructed and hence no multiplexer is used.

EXAMPLE 11.11 Design ALU of Example 11.10 without using multiplexer.

Solution: After carefully seeing function table of ALU and arithmetic circuit design in the above figure, it may be noted that

  (i) S_0 is connected to C_0 (carry input) of FA for LSB.
  (ii) Carry has got no significance in case of logical operations. Hence, it may make 0 to make the design simple. In order to make carry 0 for all logical operation arithmetic circuit may be modified using

$$C_i = S_3' C_i.$$

For example $\quad C_0 = S_3' C_0.$

So first change required is

$$C_i = S_3' C_i \qquad \ldots(1)$$

  (iii) After making carry 0 in arithmetic circuit some logical operations will be performed by arithmetic circuit which may be desired or may not be desired. Let us see what logical operation is obtained by making carry 0

S_2	S_1	Function Obtained
0	0	A EX-OR 0 EX-OR 0 (as Y is 0 when $S_2S_1 = 00$) = A EX-OR 0 = A
0	1	A EX-OR B (as Y is B when $S_2S_1 = 01$)
1	0	A EX-OR B' (as Y is B' when $S_2S_1 = 10$) = $A.B + A'.B' = A$ EX-NOR B
1	1	A EX-OR 1(Y is 1 when $S_2S_1 = 11$) = $A.0 + A'.1 = A'$

After seeing the obtained logical function it is clear that arithmetic circuit requires modification/changes when S_2S_1 are r 00 and 10 because as per ALU function table OR and AND are required, respectively. Now the questions are

  (i) How to make A as A OR B when $S_2S_1 = 00$ and
  (ii) how to make A EX-OR B' being obtained at the output of ALU into A AND B. Let us answer these questions one by one.

While answering first question all other functions should not change. In other words X should remain A for other values of control signal but for logical operation (i.e. when $S_3 = 1$) and $S_2S_1 = 00$ X should be A OR B, i.e. Boolean expression $A + B$, so Boolean expression for X should be as given below.

$$X = A + S_3' S_2' S_1' B \qquad \ldots(2)$$

In order to transform A EX-OR B' into A AND B (which is a Boolean expression $A.B$), let us assume that input A is changed to some Boolean expression say $A + M$. Then the function obtained will be $(A + M)$ EX-OR B'. Therefore,

Output of ALU = $(A + M)$ EX-OR $B' = (A + M) B + (A + M)'B' = A.B + M.B + A'.M'.B' = A.B$ if $M.B + A'.M'.B'$ get vanished.

$M.B$ should vanish when M is chosen to be B (as $B.B' = 0$ in Boolean algebra).

If M is B', then M' will be B (as complement of complement restore the original one) and so $A'.M'.B' = A'.B.B' = 0$ and hence last term $A'.M'.B'$ will also vanish.

Therefore, output of ALU $= A.B = A$ AND B if $M = B'$. It means for $S_3S_2S_1 = 110$ A should be transformed into Boolean expression $A + B$, but this transformation should not change functions for other values of control signals.

Hence using Eq. (2)

$$X = A + S_3'S_2'S_1'B + S_3S_2S_1'B' \qquad \ldots(3)$$

So, arithmetic circuit should be modified in accordance with Eqs. (1) and (3).

Hence final Boolean expression for X, Y and C_i should be as given below and the circuit may be drawn accordingly.

(i) $C_i = S_3'C_i$
(ii) $X = A + S_3'S_2'S_1'B + S_3S_2S_1'B'$
(iii) $Y = S_1B + S_2B'$

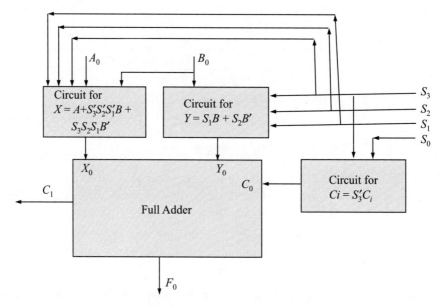

Now connect four 1-bit ALU in cascade to get 4-bit ALU without using multiplexer. N-bit ALU will be constructed by connecting N 1-bit ALU in cascade.

11.3 CENTRAL PROCESSING UNIT (CPU)

The ALU designed in the previous section may be upgraded to CPU/Processor Unit by providing Status register, Shifter register, Set of general purpose register, register selection logic for first operand A of ALU, register selection logic for second operand B of ALU and selection logic for destination register.

Status register stores the value of flags after arithmetic/logic operation performed by ALU. When carry is generated carry flag C is set. When output of ALU becomes 0 then the 0 flag Z is set. When MSB of ALU output is one sign flag S is set. When the result of computation by ALU exceeds then its range overflow flag V is set. Combinational circuit showing setting of these flags is shown in Figures. 11.7–11.9.

Sometimes output of ALU needs to be shifted. For this purpose a shifter is connected at the output of ALU that perform shifting or other operation like transfer or clear in accordance with its control signals. Let us design 4-bit shifter whose function table is given in Table 11.6.

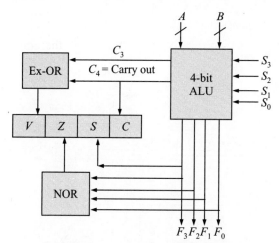

Table 11.6 Register function table

H_1	H_0	Function
0	0	No shifting
0	1	Shift left
1	0	Shift right
1	1	Clear to 0

Figure 11.7 ALU with status register.

This shifter can be designed using multiplexer. Shifter inputs are F_0, F_1, F_2, F_3 as data input and H_1, H_0 as control inputs. S_0, S_1, S_2 and S_3 are outputs.

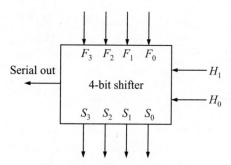

Figure 11.8 Shift block diagram.

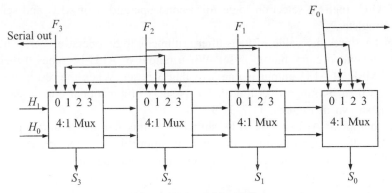

Figure 11.9 4 bits shifter.

A general register organization or bus organization for 2-bit ALU-shifter called execution unit may be designed as illustrated in Figure 11.10 for 3 one-bit registers namely X, Y, Z.

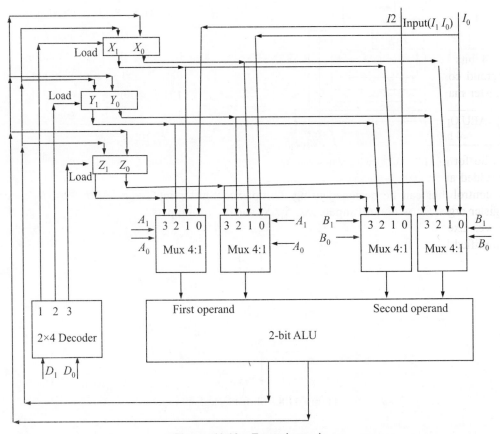

Figure 11.10 Execution unit.

The same may be extended for *n*-bit registers by increasing number of multiplexers. A register may designed with control inputs load and clr such that when load is one whatever is available at the input is loaded into the register and when clr is one register value becomes zero. The ALU along with shifter and register may be called execution unit or data processor. Execution unit may be controlled by controlling the control signals for ALU (S_3, S_2, S_1, S_0), shifter (H_1, H_0), multiplexers for first operand (A_1, A_0), second operand (B_1, B_0) and destination register (D_1, D_0).

Binary encoding of registers for first operand, second operand and destination is taken as given in Table 11.7.

Table 11.7 Register binary encoding

Binary code	First operand A_1A_0	Second operand B_1B_0	Destination D_1D_0
00	Available Input	Available input	None
01	X	X	X
10	Y	Y	Y
11	Z	Z	Z

4 bits control signals for ALU, 2 bits control signals for shifter control, 2 bits for first operand control signals, 2 bits for second operand controls signals and 2 bits for destination register may be combined into one word called control word. So, control word size is 12 bits.

ALU Operation 4 bits	Shifter Operation 2 bits	First operand 2 bits	Second operand 2 bits	Destination 2 bits

So for executing instruction ADD X, Y, Z ($X \leftarrow Y + Z$, i.e. content of registers Y and Z will be added and the sum should be put into X) control word 001000101101 should be generated by control unit and the control signals are sent to control points of ALU shifter with general register organization through wires connected between control unit and execution unit designed above. Execution unit equipped with control unit and fetch unit is central processing unit. Simplified block diagram of CPU is given in Figure 11.11.

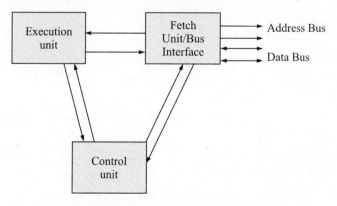

Figure 11.11 CPU simplified block diagram.

Fetch unit is responsible fetching instruction and data from memory. Bus interface unit performs the job of storing data into the memory besides fetching data and operand from memory. Instructions from memory are fetched one by one by fetch unit and transferred to control unit. Control unit decodes and interprets the instruction and accordingly generates the control signals for execution unit and also for fetch unit if operand is to be fetched from memory. Then execution unit executes the instruction as per control signals issued by control unit. Finally, results are routed to memory if required as per requirement of instruction being executed with help of controls signals generated by control unit in sequence. Fetch unit has special register called program counter or instruction pointer for keeping track of next instruction to be executed. Control unit has also a dedicated register called instruction register that contains instruction being executed. Control unit ensures the fetching of instruction, decoding of the instruction, operand fetching if required, execution and rerouting occurs in the sequence to complete execution of an instruction successfully. Power of the CPU may be improved by increasing the number of register and ALU functionalities as per instruction set decided for CPU to be designed. Set of wires used for carrying data, address and control signals are called data bus, address bus and control bus, respectively. CPU interfaced with memory and input/output devices or interfaces is called computer. CPU on a single chip is called microprocessor. Microprocessor connected with memory and input/output devices/interfaces and other peripherals is called microcomputer. In other words, computer using microprocessor as CPU is called microcomputer.

EXERCISES

Short Answer Questions

1. What is T-gate?
2. Is it possible to implement conventional logic circuit T-gate only?
3. What is weighted sum?
4. What is relationship between weighted sum and T?
5. If $w = 2a - 1b + 5c$, determine the inputs variables and weight associated with each variable.
6. What is threshold logic circuit?
7. Is threshold gate and T-gate same?
8. What is arithmetic circuit?
9. What is logic circuit?
10. What is CPU?
11. How many minterms will be generated for performing just two operation using 4-bit arithmetic circuit?
12. Draw the block diagram of shifter.
13. What is left shift and right shift?
14. Draw the diagram of ALU with status register.
15. What is shift register?

16. How many multiplexers are required in design 4 bits shifter?
17. Define V, Z, S and C.
18. How many multiplexers are required to design 1-bit ALU?

Long Answer Questions

1. Draw the threshold logic circuit for the Boolean function $\sum(1, 2, 4, 5, 6, 7)$.
2. Find the Boolean function F from the following threshold circuit.

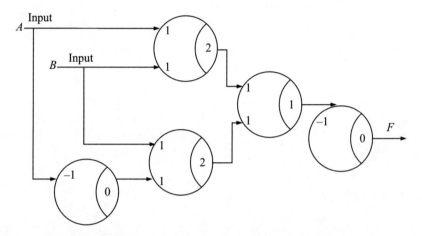

3. Draw the threshold circuit from the following I/O relation table.

| Inputs | | Weighted Sum | | Outputs | | Weighted Sum for F_1 and F_2 $w = F_1 + F_2$ | Output F |
X	Y	$T = 2$, $w = X + Y$	$T = 2$, $w = X + Y$	F_1	F_2		
0	0	0	0	0	0	0	0
0	1	1	1	0	0	0	0
1	0	1	1	0	0	0	0
1	1	2	2	1	1	2	1

4. Implement the following conventional logic circuit with T-gate.

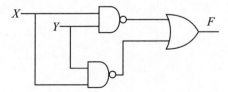

5. Implement the following Boolean function with T-gate.

$$F(A, B, C) = \prod(1, 2, 4, 5, 7)$$

6. Implement the Full Adder with threshold gate.
7. Draw threshold logic circuit from full subtractor.
8. Design a threshold logic circuit for odd parity checker.
9. Design a threshold logic circuit to convert BCD code to Excess-3 code.
10. Design a 4-bit ALU.
11. Design two bits ALU using multiplexer.
12. Design 4-bit logical circuit to perform as per function table given below.

| Control Signals | | Function |
S_2	S_1	F
0	0	A AND B
0	1	A OR B
1	0	A NOR B
1	1	NOT A

Multiple Choice Questions

1. Which of the following is relationship between weighted sum and T
   (a) Weighted sum $\geq T$
   (b) Weighted sum $\leq T$
   (c) Weighted sum $= T$
   (d) None

2. If x, y and z are three variables and weight associated the inputs are p, q and r, respectively. Which of the following is the weighted sum w?
   (a) $w = px - qy + rz$
   (b) $w = px + qy + rz$
   (c) $w = xq + yp + rz$
   (d) $w = rz + xq - px$

3. The following T-gate is equivalent to

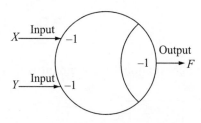

   (a) AND gate
   (b) NOR gate
   (c) NAND gate
   (d) OR gate

Threshold Circuit and Digital Computer Design   267

4. Which of the following T-gate is equivalent to OR gate?

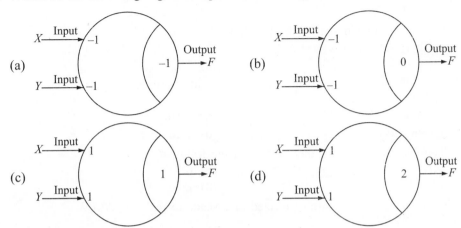

5. The weighted sum is 5 and the value of T is 6 then the output will be.
   (a) 0 (b) 1
   (c) 2 (d) 3

6. X and Y are two inputs and its associated weights are 4 and 5. If the value of T is 3 then output will be.
   (a) {0, 1, 1, 1} (b) {1, 0, 1, 1}
   (c) {1, 1, 1, 1} (d) {1, 0, 0, 1}

7. If X and Y are two inputs, output F = {0, 1, 1, 1} and T = 3 the weight associated with X and Y are
   (a) 0, 1 (b) 2, 1
   (c) 4, 5 (d) None

8. Which of the following conventional gate equation is equal to the given T-gate?

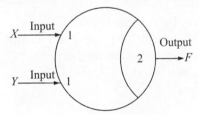

   (a) AND (b) OR
   (c) NAND (d) NOR

9. The weighted sum in the following T-gate is

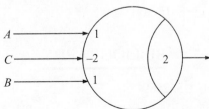

(a) $A + B - 2C$
(b) $A - B + 2C$
(c) $A - B - 2C$
(d) $A - 2B + C$

10. The set of weighted sum in the following T-gate is
    (a) {0, 1, 1, 3, 5, 6, 2}
    (b) {0, –2, 1, –1, 1, –1, 2, 0}
    (c) {0, 1, 1, 3, 5, –2, 3}
    (d) {1, 3, 1, 0, 5, 6, 2}

11. _____ register stores the value of flags after arithmetic/logic operation performed by ALU.
    (a) Status
    (b) MBR
    (c) General
    (d) Temporary

12. When carry is generated then, which of the following flag is set?
    (a) V
    (b) C
    (c) S
    (d) Z

13. How many 1-bit ALU are connected in cascade to get 4-bit ALU?
    (a) 2
    (b) 3
    (c) 4
    (d) 5

14. Which of the following is correct for the following circuit?

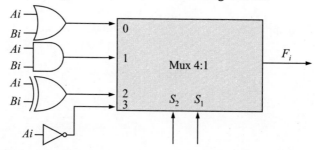

(a) 2 arithmetic circuit
(b) 2-bit logic circuit
(c) 4-bit arithmetic circuit
(d) 1-bit logic circuit

15. CPU (Central Processing Unit) consists of
    (a) ALU (Arithmetic Logic Unit), Registers and Control Unit
    (b) ALU (Arithmetic Logic Unit) and Registers
    (c) ALU (Arithmetic Logic Unit) and Control Unit
    (d) Registers and Control Unit

Answers

1. (a)	2. (b)	3. (c)	4. (c)	5. (a)	6. (a)
7. (c)	8. (a)	9. (a)	10. (b)	11. (a)	12. (b)
13. (c)	14. (d)	15. (a)			

BIBLIOGRAPHY

http://www.google.com/patents/US5256918.

Kumar, Anand., *Fundamentals of Digital Circuits*, 3rd ed., PHI Learning, 2014.

Appendix

Miscellaneous Objective Type Questions

Q1. If $(12x)_3 = (123)_x$, then the value of x is
   (a) 3   (b) 3 or 4
   (c) 2   (d) none of these

Q2. If $(x567)_8 + (2yx5)_8 = (71yx)_8$, then the values of x and y are
   (a) 4, 3   (b) 3, 3
   (c) 4, 4   (d) 4, 5

Q3. The number of 1's in the binary representation of the value of the decimal expression $16^3 \times 9 + 16^2 \times 7 + 16 \times 5 + 3$ is
   (a) 15   (b) 12
   (c) 9   (d) 6

Q4. The range of the number that can be represented in 8 bit using 2's complement representation is
   (a) −128 to +128   (b) −128 to +127
   (c) −127 to +128   (d) −127 to +127

Q5. How many 2-input multiplexers are required to construct a 1024-input multiplexer?
   (a) 1023   (b) 31
   (c) 10   (d) 127

Q6. n flip flops will divide the clock frequency by a factor of
   (a) n^2 (b) n
   (c) $2n$ (d) $\log(n)$

Q7. Three main components of a digital computer system are
   (a) memory, I/O, DMA   (b) ALU, CPU, memory
   (c) memory, CPU, I/O   (d) control circuits, ALU, registers

Q8. FFFF will be the last memory location in a memory of size
   (a) 1 K   (b) 16 K
   (c) 32 K   (d) 64 K

Q9. Three main components of a digital computer system are
  (a) memory, I/O, DMA
  (b) ALU, CPU, memory
  (c) memory, CPU, I/O
  (d) control circuits, ALU, Registers

Q10. Hamming distance between 001111 and 010011 is
  (a) 1
  (b) 2
  (c) 3
  (d) 4

Q11. Number of flip flops required to divide a given frequency by 16 is
  (a) 2
  (b) 3
  (c) 4
  (d) 5

Q12. Number of states generated by 5 flip flop is
  (a) 64
  (b) 32
  (c) 16
  (d) 8

Q13. JK flip flop may be converted into D flip flop by
  (a) shorting J and K
  (b) connecting inverted J to K
  (c) connecting inverted K to J
  (d) connecting EX-OR of J and K to J

Q14. JK flip flop may be converted into T flip flop by
  (a) shorting J and K
  (b) connecting inverted J to K
  (c) connecting inverted K to J
  (d) connecting EX-OR of J and K to J

Q15. Sum output of half adder is obtained by which of the following gates
  (a) EX-OR
  (b) AND
  (c) OR
  (d) NOT

Q16. Carry output of half adder is obtained by which of the following gates
  (a) EX-OR
  (b) AND
  (c) OR
  (d) NOT

Q17. Sum output of full adder with inputs a, b and c is obtained by which of the following Boolean expression
  (a) a EX-OR b EX-OR c
  (b) $ab + a$ EX-OR c
  (c) $ab + bc + ca$
  (d) $ab' + ab + c$

Q18. Carry output of full adder is obtained by which of the following Boolean expression
  (a) a EX-OR b EX-OR c
  (b) $ab + a$ EX-OR c
  (c) $ab + bc + ca$
  (d) $ab' + ab + c$

Q19. 3 × 8 decoder may be constructed using
  (a) two 2 × 4 decoder only
  (b) two 2 × 4 decoder and one NOT gate
  (c) two 2 × 4 decoder and two NOT gate
  (d) two 2 × 4 decoder and one OR gate

Q20. 4 × 16 decoder may be constructed using
  (a) two 3 × 8 decoders only
  (b) two 3 × 8 decoders and one NOT gate
  (c) two 3 × 8 decoder and two NOT gate
  (d) two 3 × 8 decoder and one OR gate

Q21. How many 2 × 4 decoders will be required to construct 4 × 16 decoder?
(a) 4
(b) 5
(c) 3
(d) 6

Q22. How many 2:1 multiplexer will be required to construct 4:1 multiplexer?
(a) 3
(b) 2
(c) 4
(d) 5

Q23. How many 4:1 multiplexer will be required to construct 16:1 multiplexer?
(a) 3
(b) 2
(c) 4
(d) 5

Q24. Minimum number of NAND gate required to implement exclusive OR gate is
(a) 2
(b) 3
(c) 4
(d) 5

Q25. Minimum number of NOR gates required to implement exclusive NOR gate is
(a) 2
(b) 3
(c) 4
(d) 5

Q26. Simplified Boolean expression for the function $F(a, b, c, d) = \Sigma(1, 2, 4, 7, 8, 11, 13, 14)$ is
(a) a EX-OR b EX-OR c EX-OR d
(b) a EX-NOR b EX-NOR c EX-NOR d
(c) $ab' + bc + cd$
(d) $ac' + bc' + c'd$

Q27. Simplified Boolean expression for the function $F(a, b, c, d) = \Sigma(0, 3, 5, 6, 9, 10, 12, 15)$ is
(a) a EX-OR b EX-OR c EX-OR d
(b) a EX-NOR b EX-NOR c EX-NOR d
(c) $ab' + bc + cd$
(d) $ac' + bc' + c'd$

Q28. A demultiplexer is
(a) decoder with enable input
(b) decoder without enable input
(c) multiplexer without enable input
(d) multiplexer with enable input

Q29. Any function can be implemented using
(a) decoder and OR gate
(b) decoder and Not gate
(c) decoder and AND gate
(d) decoder only

Q30. Which one of the following is called data selector?
(a) multiplexer
(b) decoder
(c) demultiplexer
(d) multiplier

Q31. Minimum number of AND gates required in the implementation of 8:1 multiplexer is
(a) 6
(b) 4
(c) 8
(d) 16

Q32. Minimum number of decoder required in the implementation of 8:1 multiplexer is
(a) 1
(b) 2
(c) 3
(d) 4

Q33. A combinational circuit obtained by giving all output lines of a decoder into the input of an OR gate is called
(a) Demultiplexer
(b) Multiplexer
(c) Decoder
(d) Multiplier

Q34. In order to implement 8 variable Boolean function which one of the following multiplexer should be used
(a) 256:1 (b) 512:1
(c) 64:1 (d) 32:1

Q35. In order to implement $f(a, b, c) = \sum(1, 3, 5, 6)$ using 4:1 multiplexer, a and b may be connected to the select input of multiplexer and I_0, I_1, I_2 and I_3 should be connected to
(a) c, c, c, c' (b) c', c, c, c'
(c) c, c', c, c (d) c, c', c', c

Q36. Which one of the following multiplexer can be used to implement 4 input exclusive OR gate
(a) 8:1 (b) 16:1
(c) 4:1 (d) 2:1

Q37. One bit adder may be made to work as adder/subtractor by adding which one of the following gate
(a) Exclusive OR gate (b) Exclusive NOR gate
(c) AND gate (d) OR gate

Q38. Size of mantisa in bits in IEEE 754 standard for representing single precision floating point number is
(a) 22 (b) 21
(c) 23 (d) 24

Q39. Size of mantisa in bits in IEEE 754 standard for representing double precision floating point number is
(a) 52 (b) 51
(c) 53 (d) 54

Q40. Size of exponent in bits in IEEE 754 standard for representing single precision floating point number is
(a) 8 (b) 7
(c) 9 (d) 10

Q41. Size of exponent in bits in IEEE 754 standard for representing double precision floating point number is
(a) 12 (b) 11
(c) 10 (d) 13

Q42. Value of bias in decimal in IEEE 754 standard for representing single precision floating point number is
(a) 128 (b) 127
(c) 129 (d) 130

Q43. Value of bias in decimal in IEEE 754 standard for representing double precision floating point number is
(a) 1022 (b) 1023
(c) 1021 (d) 1024

Q44. Range of 8 bit integers represented using 2's complement representation is
(a) −127 to +127 (b) −128 to +127
(c) −127 to +128 (d) −128 to +128

Q45. Range of 8 bit integers represented using 1's complement representation is
(a) −127 to +127
(b) −128 to +127
(c) −127 to +128
(d) −128 to +128

Q46. Characteristic equation of JK flip flop is given by
(a) $q+ = jq' + k'q$
(b) $q+ = j'q' + k'q$
(c) $q+ = jq' + k'q'$
(d) $q+ = jq + k'q$

Q47. Characteristic equation of T flip flop is given by
(a) $q+ =$ T EX-OR q
(b) $q+ =$ T AND q
(c) $q+ =$ T OR q
(d) $q+ =$ T NAND q

Q48. Characteristic equation of D flip flop is given by
(a) $q+ = D$
(b) $q+ = D$ AND q
(c) $q+ = D$ OR q
(d) $q+ =$ DEX-OR q

Q49. Which of the following T-gate represents given I/O relation table

Inputs	Weighted sum	Output
X	w = −X	F
0	0	0
1	−1	1

(a) X —Input→ [1, 0] →Output F

(b) X —Input→ [1], Y —Input→ [1], threshold 1 →Output F

(c) X —Input→ [−1], Y —Input→ [−1], threshold 0 →Output F

(d) X —Input→ [−1], Y —Input→ [−1], threshold −1 →Output F

Index

1-bit ALU, 257, 258
1-bit arithmetic circuit, 10, 252, 258
2-bit ALU-shifter, 262
2 bits control signals, 263
4-bit ALU, 258
4-bit arithmetic circuit, 252
4-bit binary counter, 218
4 bits control signals, 263
4-bit shifter, 261
4-stage Johnson ring counters, 223
4 1 multiplexer, 144
4 8 RAM, 200
4 16 decoder, 169
7's complement, 70
8-bit ASCII code, 56
8's complement subtraction, 71
8's complement subtraction (nonal number), 84
8 1 multiplexer, 146, 147
9's complement subtraction, 73
9's complement subtraction (nonal number), 85
10's complement subtraction, 74
15's complement subtraction, 76
16-bits binary number, 45
16's complement subtraction, 77, 81
32 Bits Binary Number, 43
2421 Code, 55
8$\overline{4}$21 code, 55
8421 Code, 54

Adding A Parallel Load Operation, 234
Algorithm for simplifying Boolean function using K-map, 123

Algorithmic State Machine (ASM), 200
Algorithm level, 5
ALU (Arithmetic Logic Unit), 251, 257
ALU designed, 260
Analogous, 175
Analytical engine, 3
AND gate, 96
AND gate implementation with T-gate, 247
Applications of counter, 209
Arithmetic circuit, 252, 258
Arithmetic circuit design, 251
Arithmetic operations, 3
ASCII code, 56
ASM chart method, 201
Asynchronous, 209
Asynchronous counter, 226

BCD Code, 52
BCD decoder, 158
Bidirectional Shift Register (BSR), 235, 240
Binary addition, 65, 69
Binary arithmetic, 65
Binary counter, 210
Binary digits, 41
Binary fraction number, 29
Binary fraction number to decimal conversion, 17
Binary fraction to hexadecimal conversion, 31
Binary fraction to octal conversion, 24
Binary multiplication, 69
Binary number, 11
Binary subtraction, 66

Index

Binary systems, 92
Binary to decimal conversion, 16
Binary to gray code conversion, 53
Binary to hexadecimal conversion, 31
Binary to octal conversion, 24
Binary value, 59
Bit information, 180
Block diagram of 4 8, 200
Block diagram of clocked flip flop, 183
Block diagram of combinational circuit, 134
Block diagram of demultiplexer, 155
Block diagram of flip flops, 180
Block diagram of multiplexer, 143
Block diagram of ROM, 169
Block diagram of t type flip flop, 186
Boolean algebra, 92, 119
Boolean algebra method, 119
Boolean domain, 94
Boolean function, 94, 101, 119
Boolean function of s and c, 136
Boolean logic, 91
Boolean logic computer, 96
Boolean operators, 91
Borrow, 84
Broad-spectrum structure, 175
Bus organization, 262

CAD systems, 6
CAD tools, 6
Canonical forms, 120
Carry, 65
Cascade form, 221
CEDAR Logic Simulator, 8
Central processing unit, 263
Characteristic internal sequence, 235
Circuit, 180
Circuit diagram of half adder, 136
Circuit full Adder (FA), 252
Circuit level network, 5
Circuits diagram software, 8
Clock cycle, 209
Clocked digital systems, 3
Clock input electronic equipment, 184
Clock pulse (cp), 209, 236, 234
Code, 55
Codes and its conversion, 52

Combinational circuit, 134
Combinational Circuit Implementation with ROM, 170
Compare the coefficient, 187
Comparing coefficients, 189
Complex Boolean function, 121
Computer deals, 65
Conditional box, 201
constant element, 94
Construct 1-bit arithmetic circuit, 254
Construction RS flip flop with NAND, 182
Controlling, 263
Control signal, 143, 144, 151, 252
Control unit, 264
Conventional circuit, 250
Conventional gates, 244
Conversion from decimal to hexadecimal, 25
Conversion of decimal to hexal number, 32
Conversion of decimal to nonal number, 36
Conversion of decimal to octal number, 18
Counter, 209
Counter REST, 218
CPU (Central Processing Unit) Processor, 251, 260
Currents, 65
Cyclic cover problem, 128

Data, 41, 238
Data coming, 237
Data register (DR), 234
Data variables, 153
Decade counter, 218
Decimal arithmetic, 73
Decimal decoder, 158
Decimal equivalent, 127
Decimal fraction to hexadecimal conversion, 26
Decimal fraction to hexal conversion, 32
Decimal fraction to nonal conversion, 36
Decimal fraction to octal conversion, 18
Decimal point, 41
Decimal to binary conversion, 14
Decision box, 201
Decoder, 155
Decrypt, 220
Demultiplexer, 155
Design a combinational circuit, 134

Design procedure in case of state diagram, 193
Design procedure in case of state equation, 187
Design procedure in case of state table, 190
Detect a single error, 57
Devices/interfaces, 264
Diagram for implementation of function, 148
Diagram of CAD system, 6
Diagram of Parallel In Parallel Out (PIPO) shift register, 239
Diagram of Parallel in Serial Out, 238
Diagram of register with parallel load, 235
Difference and borrow, 139
Different digits, 65
Different types of multiplexer, 150
Digital, 1
Digital circuit, 3
Digital computer, 3
Digital electronic, 2
Digital signal processing, 2
Digital system, 2
Direct method, 67
Disadvantage of a half subtractor, 140
D type flip flop, 185

Eight's complement, 49
Eight's complement subtraction, 71
Electrical disturbance, 56
Electronic Design Automation (EDA or ECAD), 7
Electronic form, 56
Elements of set, 92
Empty set, 92
Encoder, 165
Eq designing, 166, 167
Error, 56
Error correction technique, 56
Error detection, 56
Error detection and correction code, 56
Error position, 60
Error position in message, 59
Even message, 59
Even parity, 56
Even parity method, 56, 57
Examples of combinational circuit, 134
Excess-3 code, 53
Excitation table, 191, 199, 216, 226

Exclusive-NOR gate, 100
Executing instruction, 263
Execution unit, 262, 263
Exponent part, 14, 43
Extra output logic, 175

Faulty designs, 3
Fifteen (F)'s complement, 51
Finding few prime implicants as possible, 128
Finite state machines, 200
Five variable K-map, 122
Flags, 261
Flip flop, 180
Floating number, 43
Floating Point Decimal Number to Binary Number, 14
Floating point number, 43
Floating point representation, 43
Floating Point Representation of 16 Bits Binary Number, 44
Flow of electricity, 96
Four-bit Parallel Adder, 139
Four-Bit Parallel Subtractor, 142
Four-bit register, 233
Four variable K-map, 122
Four variables function is implemented with 4 1 multiplexer, 151
Full adder, 137
Full subtractor, 140
Function, 254
Functional elements, 3
Function obtained, 259
Function table, 254

General diagram of encoder, 165
General Purpose Register (GPR), 234
General purpose register, register selection logic, 260
General register organization, 262
General term, 92
George Boole, 92
Gray Code, 52
Gray code to binary conversion, 53
GT designing, 167, 168

Half adder, 136

Half subtractor, 139
Hamming code method, 57
Hexadecimal addition, 75
Hexadecimal arithmetic, 75
Hexadecimal fraction number, 29
Hexadecimal fraction to binary conversion, 29
Hexadecimal fraction to decimal conversion, 27
Hexadecimal non-fraction, 30
Hexadecimal number, 12
Hexadecimal subtraction, 76
Hexadecimal to binary conversion, 28
Hexadecimal to decimal conversion, 27
Hexal addition, 78
Hexal arithmetic, 78
Hexal division, 80
Hexal fraction, 35
Hexal fraction to decimal conversion, 35
Hexal multiplication, 79
Hexal number, 13, 35
Hexal number to decimal conversion, 35
Hexal subtraction, 79
Hold state, 183

Implementation circuit with rom, 171
Implementation of conventional gates with t-gate, 247
Indeterminate, 183
Information, 41
Information sources, 143
Input combination, 181
Input/output, 264
Input/output (I/O) relation, 245
Instruction Register (IR), 234
Interconnected, 5
Invalid output, 184
Inverters, 142
I/O relation table, 249

JK flip flop, 184
Johnson counter, 223

Karnaugh-map (K-map), 121, 150

Logical circuit diagram, 98
Logical circuit of flip flop, 182
Logical circuits perform logical operation, 257

Logical diagram of 3 8 decoder, 157
Logical diagram of D type flip flop, 185
Logical diagram of synchronous/parallel BCD counter, 217
Logical diagram of t type flip flop, 186
Logical gates, 240
Logical operations, 251
Logic circuit, 8, 258
Logic gate, 96
Logic gates, 134
LSB, 253
LT designing, 167, 168

Magnitude comparator, 166
Main memory, 180, 233
Making group, 31
Mantissa, 43
Mantissa part, 14, 43
Mechanical digital calculating machines, 3
Memory Address Register (MAR), 234
Memory Buffer Register (MBR), 234
Memory design, 180
Microcomputer, 264
Microprocessor, 264
Minimum variable boolean function, 119
Minterm, 127
Minterms, 128
Modern application, 6
Modern processors, 233
Monolithic memories, 175
MSB, 253
Multimedia based simulator, 8
Multimedia logic digital circuit design simulator, 8
Multiplexer, 143
Multiplication performed, 79

N 1-bit arithmetic, 253
NAND gate, 98
NAND gate circuit equivalent, 102
NAND gate implementation with T-gate, 248
NAND implementation, 102
Nano number, 13
N-bit ALU, 257
N-Bit Parallel Adder, 138
N-bit registers, 263

Index

Necklace, 1
Negative number, 50
Negative number in decimal number system, 49
Next stage, 240
Nonal, 37
Nonal addition, 83
Nonal arithmetic, 83
Nonal division, 87
Nonal Fraction to Decimal Conversion, 37
Nonal multiplication, 86
Nonal number, 37, 84
Nonal number to decimal conversion, 37
Nonal subtraction, 83, 85
NOR gate, 99
NOR gate implementation with T-gate, 249
NOR implementation, 106
NOR implemented circuit, 110
Normalization, 42
Normalized, 41
NOT gate, 97
NOT gate implementation with T-gate, 248
Number systems, 11

Octal addition, 70
Octal fraction number, 21
Octal Fraction to Binary Conversion, 22
Octal Fraction to Decimal Conversion, 21
Octal number, 12, 70
Octal subtraction, 70
Octal to Binary Conversion, 22
Octal to decimal conversion, 20
Odd parity method, 56, 59
One's complement, 47
One's complement subtraction, 66
Operation in tabulation method, 127
Operator, 101
OR gate, 96
OR gate implementation with T-gate, 247
Original Boolean function, 120
Output lines, 155

PAL, 175
PAL device, 175
Parallel In Parallel Out (PIPO), 235
Parallel In Parallel Out (PIPO) Shift Register, 239
Parallel In Serial Out (PISO), 235

Parallel In Serial Out (PISO) Shift Register, 238
Parity Bit Position, 58
Partition prime implicants, 128
PCB Design Software, 8
PC (Program Counter), 209
Performing arithmetic operations, 251
Performs logic operation, 258
Perform subtraction, 79
Peripherals, 264
PLA internal logic, 172, 175
PLA (Programmable Logic Array), 172
PLA programming table, 172
Postulate, 93, 119
Power Set, 91
Present state, 191
Primary memory, 180
Procedure for NOR Gate Implementation, 107
Procedure to implement the Boolean circuit with NAND gates, 103
Processing methods, 2
Programmable ROM (PROM), 172
Proposition, 93

Quine–McCluskey, 128

Random Access Memory (RAM), 200
Range of 16 bits binary number, 45
Range of 32 bits binary number, 46
Read Only Memory (ROM), 169
Read/Write, 180
Reasoning, 93
Register, 233
Register stores four-bit data, 237
Register with parallel load, 235
Representation of negative number, 47
Representation of negative number in binary system, 47
Representation of negative number in decimal system, 49
Representation of negative number in hexadecimal system, 51
Representation of negative number in octal system, 48
Rerouting, 264
Reset, 183
Ring counter, 220

Index

Ripple counter, 209, 226
ROM truth table, 171
RS flip flop, 181
RS-latch, 181

Select, 180
Selector line, 143
Self-complement Code, 56
Self-evident, 93
Sequential circuit, 5, 180, 209
Serial In Parallel Out (SIPO), 235
Serial In Parallel Out (SIPO) Shift Register, 237
Serial In Serial Out (SISO), 235
Serial In Serial Out (SISO) Shift Register, 236
Seven's complement, 49
Seven's complement subtraction, 70
Shift registers, 235, 238
Signal, 41
Sign bit, 45
Signed magnitude, 47, 48, 50, 51
Significant, 42, 43
Simplification of Boolean function, 119
Simplified block diagram of CPU, 263
Simulating designing digital logic circuit, 8
Sixteen's complement, 52
Six variable K-map, 123
Smart connector, 8
Specification and implementation, 5
Stack Pointer (SP), 234
Stage counter cycle, 218
State Box, 201
State diagram, 210
State equation, 183, 186
State Machine diagram, 201
State table of binary counter, 215
Status register, 261
Straightforward method, 150
Subset, 92
Symbol of exclusive-NOR gate, 100

Symbol of state box, 201
Synchronous, 209
Synchronous BCD Counter, 215
Synchronous binary counter, 215
Synchronous or parallel counter, 209
Systematic process, 6

Tabulation method, 127
T flip flop, 199
T-gate, 244
Theorem, 93, 119
Theorem of Boolean algebra, 120
Threshold circuit, 246
Threshold gate, 244, 245
Threshold Gate (T-gate), 244
Threshold logic circuit, 244, 250
Threshold T, 244
Truth table, 171
Truth table of full subtractor, 140
Truth table of JK flip flop, 184
T type flip flop, 186
Two' complement subtraction, 67
Two's complement, 47, 48
Type of circuit, 96
Typical digital signal, 1

Unavailability of write operation, 169
Universally accepted, 93
Unused States, 213

Voltage, 1

Weighted code, 55
Weighted sum, 246
Weight w, 244
Writing data, 180

Xilinx and ModelSim, 7